# Hydraulic Fracturing and Safe Drinking Water Act Issues

**Mary Tiemann**
Specialist in Environmental Policy

**Adam Vann**
Legislative Attorney

April 10, 2012

Congressional Research Service

7-5700

www.crs.gov

R41760

CRS Report for Congress ────────────────────

*Prepared for Members and Committees of Congress*

# Summary

Hydraulic fracturing is a technique developed initially to stimulate oil production from wells in declining oil reservoirs. More recently, it has been used to initiate oil and gas production in unconventional (i.e., low-permeability) reservoirs where these resources were previously inaccessible. This process now is used in more than 90% of new oil and gas production wells. Hydraulic fracturing is done after a well is drilled and involves injecting large volumes of water, sand (or other propping agent), and specialized chemicals under enough pressure to fracture the formations holding the oil or gas. The sand or other proppant holds the fractures open to allow the oil or gas to flow freely out of the formation and into a production well.

Its application, along with horizontal drilling, for production of natural gas (methane) from coal beds, tight gas sands, and, more recently, from unconventional shale formations, has resulted in the marked expansion of estimated U.S. natural gas reserves in recent years. Similarly, hydraulic fracturing is enabling the development of unconventional domestic oil resources, such as the Bakken Formation in North Dakota and Montana. However, the rapidly increasing and geographically expanding use of fracturing, along with a growing number of citizen complaints and state investigations of well water contamination attributed to this practice, has led to calls for greater state and/or federal environmental regulation and oversight of this activity.

Historically, the Environmental Protection Agency (EPA) had not regulated the underground injection of fluids for hydraulic fracturing of oil or gas production wells. In 1997, the U.S. Court of Appeals for the 11th Circuit ruled that fracturing for coalbed methane (CBM) production in Alabama constituted underground injection and must be regulated under the Safe Drinking Water Act (SDWA). This ruling led EPA to study the risk that hydraulic fracturing for CBM production might pose to drinking water sources. In 2004, EPA reported that the risk was small, except where diesel was used, and that regulation was not needed. However, to address regulatory uncertainty the ruling created, the Energy Policy Act of 2005 (EPAct 2005) revised the SDWA term "underground injection" to explicitly exclude the injection of fluids and propping agents (except diesel fuel) used for hydraulic fracturing purposes. Consequently, EPA currently lacks authority under the SDWA to regulate hydraulic fracturing, except where diesel fuel is used. However, as the use of this process has grown, some in Congress would like to revisit this statutory exclusion.

Several relevant bills are pending. H.R. 1084 and S. 587 would repeal the exemption for hydraulic fracturing operations established in EPAct 2005, and amend the term "underground injection" to include explicitly the injection of fluids used in hydraulic fracturing operations, thus authorizing EPA to regulate this process under the SDWA. The bills also would require disclosure of the chemicals used in the fracturing process. S. 2248 and H.R. 4322 would specify that a state has sole authority to regulate hydraulic fracturing on federal lands within state boundaries. EPA's FY2010 appropriations act urged the agency to study the relationship between hydraulic fracturing and drinking water quality. The interim report, expected in 2012, may help inform Congress on whether federal action is needed. Meanwhile, numerous states are reviewing or have revised their oil and gas rules to address the increased use of hydraulic fracturing.

This report reviews past and proposed treatment of hydraulic fracturing under the SDWA, the principal federal statute for regulating the underground injection of fluids to protect groundwater sources of drinking water. It reviews current SDWA provisions for regulating underground injection activities, and discusses some possible implications of, and issues associated with, enactment of legislation authorizing EPA to regulate hydraulic fracturing under this statute.

# Contents

## Figures

## Tables

# Contacts

# Introduction

## Background—Hydraulic Fracturing in Oil and Gas Production

The process of hydraulic fracturing was developed initially in the 1940s to stimulate production from oil reservoirs with declining productivity. More recently, this practice has been used to initiate oil and gas production in unconventional (low-permeability) oil and gas formations.[1] Its application—in combination with technological breakthroughs, such as horizontal drilling—in the production of natural gas from coal beds, tight gas sands,[2] and unconventional shale formations has resulted in the marked expansion of estimated U.S. natural gas reserves in recent years. Similarly, hydraulic fracturing is enabling the development of unconventional domestic oil resources, such as the Bakken Formation in North Dakota and Montana. However, the rapidly increasing and geographically expanding use of fracturing, along with a growing number of complaints of well water contamination and other water quality problems attributed to this practice, has led to calls for greater state and/or federal oversight of this activity.

Hydraulic fracturing involves injecting into production wells large volumes of water, sand or other proppant,[3] and specialized chemicals under enough pressure to fracture low-permeability geologic formations containing oil and/or natural gas.[4] The sand or other proppant holds the new fractures open to allow the oil or gas to flow freely out of the formation and into a production well. Fracturing fluid and water remaining in the fracture zone can inhibit oil and gas production, and must be pumped back to the surface. The fracturing fluid—"flowback"—along with any naturally occurring formation water pumped to the surface, together called produced water, typically has been disposed through deep well injection or treated before disposal into surface waters.[5] According to industry estimates for various geographic areas, the volume of flowback water can range from less than 30% to more than 70% of the original fracture fluid volume.[6]

The use of hydraulic fracturing continues to increase significantly, as more easily accessible oil and gas reservoirs have declined and companies move to develop unconventional oil and gas formations. Hydraulic fracturing is used for oil and/or gas production in all 33 U.S. states where

---

[1] Hydraulic fracturing is also used for other purposes, such as developing water supply wells and geothermal production wells. This report focuses only on its use for oil and gas development.

[2] Tight gas sands are sandstone formations with very low permeability that must fractured to release the gas.

[3] According to the Schlumberger *Oilfield Glossary*, propping agents, or proppants, are "sized particles mixed with fracturing fluid to hold fractures open after a hydraulic fracturing treatment. In addition to naturally occurring sand grains, man-made or specially engineered proppants, such as resin-coated sand or high-strength ceramic materials like sintered bauxite, may also be used." The glossary is available at http://www.glossary.oilfield.slb.com/default.cfm.

[4] This process is distinct from enhanced oil and gas recovery and other secondary and tertiary hydrocarbon recovery techniques which involve separate wells. Injections for hydraulic fracturing are done through the production wells.

[5] The Schlumberger glossary notes that "produced fluid is a generic term used in a number of contexts but most commonly to describe any fluid produced from a wellbore that is not a treatment fluid. The characteristics and phase composition of a produced fluid vary and use of the term often implies an inexact or unknown composition." "Flowback" refers to "the process of allowing fluids to flow from the well following a treatment, either in preparation for a subsequent phase of treatment or in preparation for cleanup and returning the well to production."

[6] U.S. Department of Energy, Office of Fossil Energy and National Technology Laboratory, *Modern Shale Gas Development in the United States A Primer*, DE-FG26-04NT15455, April 2009, p. 66, http://fossil.energy.gov/programs/oilgas/publications/naturalgas_general/Shale_Gas_Primer_2009.pdf.

---

oil and natural gas production takes place. According to industry estimates, hydraulic fracturing has been applied to more than 1 million wells nationwide.[7]

The frequency of its use expanded markedly in the 1980s and 1990s with its application to coalbed methane (CBM) development. CBM production through wells began in the 1970s as a safety measure in coal mines to reduce the explosion hazard posed by methane. In 1984, fewer than 100 coalbed wells existed in the United States.[8] In the 1980s, demand for natural gas, new fracturing technologies, and federal tax credits for nonconventional fuels production led to dramatic growth in the CBM development industry. By 1990, nearly 8,000 coalbed wells had been drilled nationwide. In 2008, the Environmental Protection Agency (EPA) identified 56,000 CBM wells managed by operators in 692 different projects.[9]

Other unconventional gas resource formations relying on hydraulic fracturing include tight sands gas and shale gas. The Department of Energy's (DOE's) Energy Information Administration (EIA) reports that natural gas from tight sand formations is the largest source of unconventional production, while production from shale formations is the fastest growing source.[10] **Figure 1** illustrates different types of natural gas reservoirs.

The number of onshore gas wells in the United States increased from approximately 260,000 wells in 1989 to 493,100 wells in 2009.[11] According to the Independent Petroleum Association of America (IPAA), more than 90% of new natural gas wells in the United States rely on hydraulic fracturing, and together they have accounted for the production of more than 600 trillion cubic feet of gas. Similarly, fracturing is increasingly applied to U.S. oil production, and more than 7 billion barrels of oil have been produced using this process.

As noted, it is the combination of hydraulic fracturing and directional drilling that is allowing the economic development of unconventional oil and gas resources. Improvements in technology also have led to increased use of horizontal drilling in developing unconventional gas formations. Currently, shale gas production involves drilling a well vertically and then drilling horizontally out from the wellbore. Because of the low permeability of shales, more wells must be drilled into a reservoir than more permeable, conventional reservoirs. A benefit of horizontal drilling through a producing shale layer is that one well pad that utilizes horizontal well drilling can replace numerous individual well pads and reduce the surface density of wells in an area. Six to eight horizontal wells, and potentially more, can be drilled from a single well pad and access the same reservoir. According to a report prepared for DOE:

> the spacing interval for vertical wells in the gas shale plays averages 40 acres per well for initial development. The spacing interval for horizontal wells is likely to be approximately

---

[7] American Petroleum Institute, *Hydraulic Fracturing*, http://www.api.org/policy/exploration/hydraulicfracturing.

[8] U.S. Environmental Protection Agency, Study Design for Evaluation of Impacts to Underground Sources of Drinking Water by Hydraulic Fracturing of Coalbed Methane Reservoirs, http://www.epa.gov/safewater/uic/wells_coalbedmethanestudy_finalstudydesign.html.

[9] U.S. Environmental Protection Agency, *Effluent Guidelines Coalbed Methane Extraction Detailed Study*, http://water.epa.gov/scitech/wastetech/guide/cbm_index.cfm.

[10] DOE reports that proved reserves of shale gas increased from 21,735 billion cubic feet (bcf) in 2007 to 32,825 bcf in 2008. U.S. Department of Energy, Energy Information Administration, *Natural Gas Navigator Shale Gas Proved Reserves*, October 29, 2009, http://tonto.eia.doe.gov/dnav/ng/ng_enr_shalegas_s1_a.htm.

[11] U.S. Energy Information Administration, *Natural Gas Navigator Number of Producing Gas Wells*, August 2009, http://tonto.eia.doe.gov/dnav/ng/ng_prod_wells_s1_a.htm.

160 acres per well. Therefore, a 640-acre section of land could be developed with a total of 16 vertical wells, each on its own individual well pad, or by as few as 4 horizontal wells all drilled from a single multi-well drilling pad.[12]

**Figure 1. Geologic Nature of Major Sources of Natural Gas in the United States**

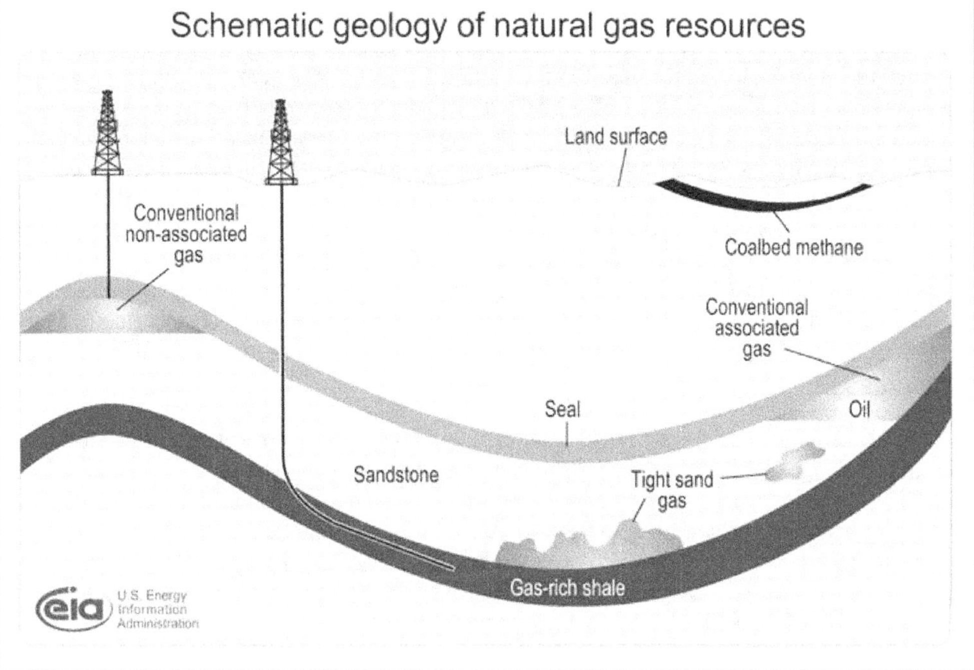

**Source:** U.S. Energy Information Administration, Independent Statistics and Analysis, October 2008. Available at http://www.eia.gov/oil_gas/natural_gas/special/ngresources/ngresources.html.

**Notes:** The diagram shows schematically the geologic nature of most major U.S. sources of natural gas:

• Gas-rich shale is the source rock for many natural gas resources, but, until [recently], has not been a focus for production. Horizontal drilling and hydrau ic fracturing have made shale gas an economically viable alternative to conventional gas resources.

• Conventional gas accumulations occur when gas migrates from gas rich shale into an overlying sandstone formation, and then becomes trapped by an overlying impermeable formation, called the seal. Associated gas accumulates in conjunction with oil, while non-associated gas does not accumulate with oil.

• Tight sand gas accumulations occur in a variety of geologic settings where gas migrates from a source rock into a sandstone formation, but is limited in its ability to migrate upward due to reduced permeabi ity in the sandstone.

• Coalbed methane does not migrate from shale, but is generated during the transformation of organic material to coal.

---

[12] Ground Water Protection Council and ALL Consulting, *Modern Shale Gas Development in the United States A Primer*, U.S. Department of Energy, Office of Fossil Energy and National Energy Technology Laboratory, April 2009, pp. 47-48, http://www.netl.doe.gov/technologies/oil-gas/publications/EPreports/Shale_Gas_Primer_2009.pdf.

A single gas production well may be fractured multiple times, using from 500,000 gallons to more than 6 million gallons of water, with compounds and proppants of various amounts added to the water.[13] Slickwater fracturing, which involves adding chemicals to increase fluid flow, is a more recent development used for unconventional shale gas development.[14]

## Hydraulic Fracturing and Drinking Water Issues

Although the rapid growth in the use of hydraulic fracturing and directional drilling to develop unconventional natural gas resources has enabled the industry to markedly expand gas production, concern has emerged regarding the potential impacts that this process may have on groundwater quality and specifically on private wells and public water supplies. The process of developing a shale gas well (drilling through an overlying aquifer, completing and casing the well, stimulating the well via hydraulic fracturing, and producing the gas) is an issue of concern for increasing the risk of groundwater contamination. During hydraulic fracturing, new fractures are induced into the shale formation, or existing fractures are lengthened. As exploration and production activities have increased and expanded into more populated areas, so has concern that the hydraulic fracturing process might introduce chemicals, methane, and other contaminants into aquifers.

Another concern involves the potential contamination of drinking water wells from surface activities. A water well that is not constructed and cased properly might be at risk if contaminated water flows from the land surface and enters the water well, possibly compromising the quality of drinking water in the well and even the aquifer itself. In such instances, and particularly where natural gas drilling and stimulation activities are nearby, leaky surface impoundments, accidental spills, or careless surface disposal of drilling fluids at the natural gas production site could increase the risk of contaminating the nearby water well.

Other water quality concerns involve the management (storage, treatment and disposal) of water produced in the fracturing process. Broader environmental issues associated with the more concentrated and geographically expanding development of unconventional gas resources include water withdrawals from streams, lakes and aquifers; potential air quality impacts; and land use changes (including those related to the development of access roads, pipelines and drill pads). Although such issues can be significant for state regulators, gas developers, local communities and landowners; they are not addressed in this report.[15]

---

[13] Multiple fractures are typical in deep shale formations. Scott Stevens and Vello Kuuskraa of Advanced Resources International report that "[t]oday, deep shale drillers all employ essentially the same Barnett-style well drilling and completion design: ±4,000-ft. long lateral stimulated by multimillion-lb slick-water fracs in a dozen stages." Source: Seven Plays Dominate North America Activity, *Oil and Gas Journal*, September 28, 2009, v. 107, n. 36 p. 41.

[14] Using slickwater fracturing increases the rate at which fluid can be pumped down the wellbore to fracture the shale. The process may involve the use of friction reducers, biocides, surfactants, and scale inhibitors. Biocides prevent bacteria from clogging wells; surfactants help keep the sand or other proppant suspended. Slickwater fracturing was first used in the Barnett shale in Texas.

[15] The scope of this report is limited to potential issues related to hydraulic fracturing and contamination of underground sources of drinking water related to the fracturing process. Another environmental concern related to hydraulic fracturing is the disposal or treatment of "flowback" from the fracturing/drilling process, which may present environmental and regulatory issues and also water treatment infrastructure issues. Disposal of flowback by means other than disposal through injection wells is regulated pursuant to the Clean Water Act. For a discussion of the hydraulic fracturing process and potential sources of water contamination, including surface water contamination, see CRS Report R42333, *Marcellus Shale Gas Development Potential and Water Management Issues and Laws*, by Mary (continued...)

Public complaints of impacts to well water have increased as gas development has intensified. In 2009, the Ground Water Protection Council (GWPC)[16] reported that several citizen complaints of well contamination attributed to hydraulic fracturing appeared to be related to hydraulic fracturing of CBM zones that were in relatively close proximity to underground sources of drinking water.[17] Additional contamination incidents in other gas producing areas have been reported.[18] In Pennsylvania, regulators confirmed that methane had migrated to water wells from drilling sites and issued notices of violations to a drilling company for, among other things, "failure to prevent gas from entering fresh groundwater."[19] Other incidents are under investigation.

EPA Region 8 is investigating the potential role of hydraulic fracturing in the contamination of a cluster of water wells in the Pavillion, WY, area. EPA began testing water wells after local residents contacted EPA in 2008 to report changes in the quality and quantity of water following nearby gas development. Under the Comprehensive Environmental Response, Cleanup, and Liability Act (CERCLA, commonly known as Superfund),[20] the residents petitioned EPA to investigate whether groundwater contamination exists, its extent, and possible sources. In December 2011, EPA issued a draft report.[21] The draft report indicated that EPA had identified certain constituents in groundwater above the production zone of the Pavillion natural gas wells that are consistent with some of the constituents used in natural gas well operations, including the process of hydraulic fracturing. EPA also stated that its approach indicates that gas production activities have likely enhanced the migration of natural gas in the aquifer and the migration of gas to domestic wells in the area. EPA did not appear to conclude that there was a definitive link to a release from the production wells. The draft report is being peer reviewed, and EPA is continuing its investigation.[22]

---

(...continued)

Tiemann et al. For a discussion of the "discharge" requirements under the Clean Water Act, see EPA, *Natural Gas Drilling in the Marcellus Shale  NPDES Program Frequently Asked Questions*, March 16, 2011, http://www.epa.gov/npdes/pubs/hydrofracturing_faq.pdf. EPA has initiated a rulemaking to control the discharge of wastewater produced by CBM and shale gas extraction. See EPA website, *Effluent Guidelines (Clean Water Act section 304(m))  2010 Effluent Guidelines Program Plan*, http://water.epa.gov/lawsregs/lawsguidance/cwa/304m/.

[16] The GWPC is a national association representing state groundwater and UIC agencies whose mission is to promote protection and conservation of groundwater resources for beneficial uses. The stated purpose of the GWPC is "to promote and ensure the use of best management practices and fair but effective laws regarding comprehensive ground water protection." http://www.gwpc.org

[17] Ground Water Protection Council, U.S. Department of Energy, Office of Fossil Energy, National Energy Technology Laboratory, *State Oil and Natural Gas Regulations Designed to Protect Water Resources*, May 2009, p. 24. Coal beds are often a source of good quality groundwater, thus, presenting challenges to developers and potential conflicts with well owners.

[18] For a discussion of environmental concerns and recommendations, see, for example, Environmental Working Group, *Drilling Around the Law*, January 2010, http://static.ewg.org/files/EWG-2009drillingaroundthelaw.pdf.

[19] Pennsylvania Department of Environmental Protection, Consent Order and Agreement, November 4, 2009; http://s3.amazonaws.com/propublica/assets/natural_gas/final_cabot_co-a.pdf.

[20] 42 U.S.C. §9605(b). This section, CERCLA §105(d), provides the authority for any person who is, or may be, affected by a release or threatened release of a hazardous substance, pollutant, or contaminant to petition the President to assess the potential hazards to public health and the environment. EPA is required to complete a preliminary assessment of a site within 12 months of the submission of a petition, or to provide an explanation of why an assessment may not be appropriate.

[21] See CRS Report R42327, *The EPA Draft Report of Groundwater Contamination Near Pavillion, Wyoming  Main Findings and Stakeholder Responses*, by Peter Folger, Mary Tiemann, and David M. Bearden.

[22] U.S. Environmental Protection Agency, Region 8, *Groundwater Investigation  Pavillion*, http://www.epa.gov/ (continued...)

---

In many cases, the source or cause of well-water contamination remains undetermined. Identifying the cause of contamination can be difficult for various reasons, including the complexity of hydrogeologic processes and investigations, a lack of baseline testing of nearby water wells prior to drilling and fracturing, as well as the confidential business information status typically given to fracturing compounds across the states. In other cases, contamination incidents have been attributed to poor well construction or surface activities, rather than the specific hydraulic fracturing process. Responding to a recent survey, major oil and gas producing states asserted that the hydraulic fracturing process has not been linked directly to groundwater contamination. However, contamination incidents attributed to poor well construction have raised concerns regarding the adequacy and/or enforcement of state well construction regulations (covering, for example, cementing, casing, and backflow prevention) for purposes of managing oil and gas development that is increasingly dependent on fracturing.

A key barrier to better understanding groundwater contamination risks that may be associated with hydraulic fracturing is the lack of scientific studies to assess the practice and related complaints. A further issue is that EPA developed the underground injection control (UIC) program primarily to regulate wells that received fluids injected for the long term or for enhanced recovery operations, but excluded oil and gas *production* wells. This distinction could raise regulatory challenges and the possibility that the agency may need to develop an essentially new framework to address hydraulic fracturing of production wells.

Such information gaps and regulatory issues contribute to uncertainty over a possible legislative or regulatory framework that might be developed for hydraulic fracturing activities under SDWA, as well as the potential costs and benefits associated with any measures. The sheer number of wells that rely on fracturing, typically multiple times, suggests that significant new resources could be required by state and federal regulators to implement and enforce any new EPA requirements on top of existing state requirements. Similarly, oil and gas industry representatives have expressed concern that well owners and operators might experience impacts, such as higher operation and compliance costs, and delays in permitting and disruption of operations, particularly early on, as any regulatory requirements are put into place and state programs are revised and then reviewed and approved by EPA (and as EPA develops regulations and implements requirements directly for nonprimacy states). States would also have to integrate any new requirements and programs with their existing oil and gas regulatory programs. The adjustments would vary among the states, reflecting different state rules and regulatory structures.

Experience regulating hydraulic fracturing in Alabama and regulatory developments in several states (including Colorado, New York, North Dakota, Pennsylvania, Ohio, Texas, West Virginia, and Wyoming) to address the growing reliance on hydraulic fracturing may add insight to the possible implications of proposed federal legislation and any subsequent regulations. Additionally, Congress has directed EPA to conduct a study on the relationship between hydraulic fracturing and drinking water.[23] The agency expects to report on the interim research results in 2012, and issue a follow-up report in 2014.

---

(...continued)

region8/superfund/wy/pavillion/index.html.

[23] P.L. 111-88, H.Rept. 111-316.

# The Safe Drinking Water Act and the Federal Role in Regulation of Underground Injection

## Review of Relevant SDWA UIC Provisions

To properly evaluate studies and any new federal or state action, it is important to understand the existing statutory and regulatory framework. Most public water systems and nearly all rural residents rely on groundwater as a source of drinking water. Because of the nationwide importance of underground sources of drinking water, Congress included groundwater protection provisions in the 1974 Safe Drinking Water Act (SDWA). The SDWA, among other things, directs the EPA to regulate the underground injection of fluids (including solids, liquids, and gases) to protect underground sources of drinking water.[24]

Part C of the SDWA establishes the national regulatory program for the protection of underground sources of drinking water, including the oversight and limitation of underground injections that could affect aquifers through the establishment of underground injection control (UIC) regulations. Key UIC requirements and exceptions contained in SDWA, Part C, include:

- Section 1421 of the SDWA directs the EPA Administrator to promulgate regulations for state UIC programs, and mandates that the EPA regulations "contain minimum requirements for programs to prevent underground injection that endangers drinking water sources." Section 1421(b)(2) specifies that EPA:

  may not prescribe requirements for state UIC programs which interfere with or impede—(A) the underground injection of brine or other fluids which are brought to the surface in connection with oil or natural gas production or natural gas storage operations, or (B) any underground injection for the secondary or tertiary recovery of oil or natural gas, *unless such requirements are essential to assure that underground sources of drinking water will not be endangered by such injection.*[25] [emphasis added]

- Section 1421(d), as amended by Energy Policy Act of 2005 (EPAct 2005),[26] specifies that the term "underground injection" as it is used in the SDWA, means the subsurface emplacement of fluids by well injection, and specifically excludes the underground injection of fluids or propping agents associated with hydraulic fracturing operations related to oil, gas, or geothermal production activities.[27] The use of diesel fuels in hydraulic fracturing, however, forfeits eligibility for this exclusion from the definition of "underground injection."[28]

- Section 1422 authorizes EPA to delegate primary enforcement authority (primacy) for UIC programs to the states, provided that the state program meets EPA requirements promulgated under Section 1421 and prohibits any

---

[24] The Safe Drinking Water Act of 1974 (P.L. 93-523) authorized the UIC program at EPA. UIC provisions are contained in SDWA Part C, §§1421-1426; 42 U.S.C. §§300h-300h-5.

[25] 42 U.S.C. §300h(b)(2).

[26] P.L. 109-58, §322.

[27] 42 U.S.C. §300h(d).

[28] *Id.*

underground injection that is not authorized by a state permit or rule.[29] If a state's UIC program plan is not approved, or the state has chosen not to assume program responsibility, then EPA must implement the UIC program in that state.

- Section 1425 authorizes EPA to approve the portion of a state's UIC program that relates to "any underground injection for the secondary or tertiary recovery of oil or natural gas" if the state program meets certain requirements of Section 1421 and represents an effective program to prevent underground injection which endangers drinking water sources.[30] Under this provision, states may demonstrate to EPA that their existing programs for oil and gas injection wells are effective in preventing endangerment of underground sources of drinking water. This provides states with an alternative to meeting the specific requirements contained in EPA regulations promulgated under Section 1421.[31] (See discussion on p. 11.)

- Section 1423 authorizes EPA enforcement actions for UIC regulatory violations.

- Section 1431 applies broadly to the SDWA and grants the EPA Administrator emergency powers to issue orders and commence civil actions to protect public water systems or underground sources of drinking water.[32]

- Section 1449, another broadly applicable SDWA provision, authorizes citizen civil actions against persons allegedly in violation of the act's enforceable requirements, or against EPA for allegedly failing to perform a duty. State-administered oil and gas programs may not have such provisions, so this could represent an expansion in the ability of citizens to challenge administration of statutes and regulations related to hydraulic fracturing and drinking water, were the hydraulic fracturing exemption provision to be repealed.

---

[29] 42 U.S.C. §300h-1. The minimum requirements for a state UIC program can be found at 40 C.F.R. Part 145.

[30] 42 U.S.C. §300h-4. SDWA Section 1425 was added by the Safe Drinking Water Act Amendments of 1980, P.L. 96-502. The House committee report accompanying the legislation that added Section 1425 noted that:

> Most of the 32 states that regulate underground injection related to the recovery or production of oil or natural gas (or both) believe they have programs already in place that meet the minimum requirements of the Act including the prevention of underground injection which endangers drinking water sources. This is especially true of the major producing states where underground injection control programs have been underway for years. It is the Committee's intent that states should be able to continue these programs unencumbered with additional Federal requirements if they demonstrate that they meet the requirements of the Act. (U.S. House of Representatives, Committee on Interstate and Foreign Commerce, *Safe Drinking Water Act Amendments*, H. Rept. 96-1348 to accompany H.R. 8117, 96th Congress, 2d Session, September 19, 1980, p. 5.)

[31] SDWA Section 1425 requires a state to demonstrate that its UIC program meets the requirements of Section 1421(b)(1)(A) through (D) and represents an effective program (including adequate record keeping and reporting) to prevent underground injection which endangers underground sources of drinking water. To receive approval under Section 1425's optional demonstration provisions, a state program must include permitting, inspection, monitoring, and record-keeping and reporting requirements.

[32] 42 U.S.C. §300i. The Administrator may take action when information is received that (1) a contaminant is present in or is likely to enter a public drinking water supply system or underground source of drinking water "which may present an imminent and substantial endangerment to the health of persons," and (2) the appropriate state or local officials have not taken adequate action to protect such persons.

## The "Endangerment" Standard

As noted, the SDWA states that UIC regulations must "contain minimum requirements for effective programs to prevent underground injection which endangers drinking water sources."[33] Known as the "endangerment standard," this statutory standard is a major driving force in EPA regulation of underground injection.

The endangerment language focuses on protecting groundwater that is used or may be used to supply public water systems. This focus parallels the general scope of the statute, which addresses the quality of water provided by public water systems and does not address private, residential wells. The endangerment language has raised questions as to whether EPA regulations can reach underground injection activities to protect groundwater that is not used by public water systems.

## Defining "Underground Source of Drinking Water"

The SDWA directs EPA to protect against endangerment of an "underground source of drinking water" (USDW). The statute defines a USDW to mean an aquifer or part of an aquifer that either:

- supplies a public water system, or

- contains a sufficient quantity of groundwater to supply a public water system;[34] and

  - currently supplies drinking water for human consumption; or

  - contains fewer than 10,000 milligrams per liter (mg/L) total dissolved solids; and

- is not an "exempted aquifer."[35]

In a 2004 report on hydraulic fracturing of coalbed methane reservoirs, the agency further noted that the "EPA also assumes that all aquifers contain sufficient quantity of groundwater to supply a public water system, unless proven otherwise through empirical data."[36] However, because these expanded agency characterizations of what constitutes a USDW are not included in SDWA or related regulation, and, therefore, are not binding on the agency, it is uncertain how they might be

---

[33] 42 U.S.C. §300h(b)(1).

[34] EPA further explained this requirement in a 1993 memorandum which provided that "[t]o better quantify the definition of USDW, EPA determined that any aquifer yielding more than 1 gallon per minute can be expected to provide sufficient quantity of water to serve a public water system and therefore falls under the definition of a USDW." EPA Memorandum: *Assistance on Compliance of 40 CFR Part 191 with Ground Water Protection Standards.* From James R. Elder, Director, Office of Ground Water and Drinking Water, to Margo T. Oge, Director, Office of Radiation and Indoor Air, June 4, 1993.

[35] §40 C.F.R. 144.3. According to EPA regulations, an exempted aquifer is an aquifer, or a portion of an aquifer, that meets the criteria for a USDW, for which protection has been waived under the UIC program. Under 40 C.F.R. Part 146.4, an aquifer may be exempted if it is not currently being used—and will not be used in the future—as a drinking water source, or it is not reasonably expected to supply a public water system due to a high total dissolved solids content. The SDWA does not mention aquifer exemption, but EPA explains that without aquifer exemptions, certain types of energy production, mining, or waste disposal into USDWs would be prohibited. EPA, typically at the Region level, makes the final determination on granting all exemptions.

[36] U.S. Environmental Protection Agency, *Evaluation of Impacts to Underground Sources of Drinking Water by Hydraulic Fracturing of Coalbed Methane Reservoirs*, EPA 816-R-04-003, June 2004, pp. 1-5.

---

applied in future situations. Notably, the SDWA does not prohibit states from establishing requirements that are stricter than federal requirements, and many states have their own definitions and classifications for groundwater resources.

## Underground Injection Control Regulatory Program Overview

To implement the UIC program as mandated by the provisions of the SDWA described above, EPA has established six classes of underground injection wells based on categories of materials that are injected by each class into the ground. In addition to the similarity of fluids injected in each class of wells, each class shares similar construction, injection depth, design, and operating techniques. The wells within a class are required to meet a set of appropriate performance criteria for protecting underground sources of drinking water (USDW). The six well categories are briefly described below, including the estimated number of wells nationwide.[37]

- Class I wells inject hazardous wastes, industrial non-hazardous liquids, or municipal wastewater beneath the lowermost USDW. (There are 650 such wells regulated as Class I wells in the United States.) The most stringent regulations apply to these wells.

- Class II wells inject brines and other and fluids associated with oil and gas production, and hydrocarbons for storage. The wells inject fluids beneath the lowermost USDW (151,000 wells). Section 1425, which allows states to apply their own regulations in lieu of EPA regulations, applies to Class II wells.

- Class III wells inject fluids associated with solution mining of minerals (e.g., salt and uranium) beneath the lowermost USDW (21,400 wells).

- Class IV wells inject hazardous or radioactive wastes into or above USDWs. These wells are banned unless authorized under a federal or state groundwater remediation project (24 wells).

- Class V includes all injection wells not included in Classes I-IV, including experimental wells. Class V wells frequently inject non-hazardous fluids into or above USDWs and are typically shallow, on-site disposal systems. However, some deep Class V wells inject below USDWs (500,000-650,000).[38]

- Class VI wells: In 2010, EPA issued a rule for Class VI wells, to be used for the geologic sequestration of carbon dioxide (0).

The UIC regulatory program includes the following broad elements: site characterization, area of review, well construction, well operation, site monitoring, well plugging and post-injection site care, public participation, and financial responsibility. While the six classes broadly share similar regulatory requirements, those for Class I wells are the most comprehensive and stringent. **Table 1** outlines the shared minimum technical requirements for Class I, II, and III wells.

---

[37] Regulatory requirements for state UIC programs are established in 40 C.F.R. §§144-147.

[38] U.S. Environmental Protection Agency, Underground Injection Control Program, Classes of Wells, http://water.epa.gov/type/groundwater/uic/wells.cfm. The inventory of Class V wells is incomplete.

---

**Table 1. Minimum Federal Technical Requirements for Class I, II, and III Wells**

**Permitting Requirements Common to Class I, II, and III Wells**

Demonstration that casing and cementing are adequate to prevent movement of fluid into or between USDWs. Cement bond logs are often needed to evaluate/verify the adequacy of the cementing records.

Financial assurances (bond, letter of credit, or other adequate assurance) that the owner or operator will maintain financial responsibility to properly plug and abandon the wells.

A maximum operating pressure calculated to avoid initiating and/or propagating fractures that would allow fluid movement into a USDW.

Monitoring and reporting requirements.

Requirement that all permitted (and rule authorized) wells which fail mechanical integrity be shut in immediately. A well may not resume injection until mechanical integrity has been demonstrated.

Schedule for demonstrating mechanical integrity (at least every five years for Class I nonhazardous, Class II, and Class III salt recovery wells).[a]

All permitted injection wells, which have had the tubing disturbed, must have a pressure test to demonstrate mechanical integrity.

Plans for plugging and abandonment. All Class I, II, and III wells must be plugged with cement.

**Source:** U.S. EPA Technical Program Overview: Underground Injection Control Regulations, EPA 816-R-02-025, December 2002, p. 65.

a. Class I hazardous wells must demonstrate mechanical integrity once a year.

## Class II Wells

Because this discussion of hydraulic fracturing is related to oil and gas production, this report focuses primarily on regulatory requirements for Class II wells rather than other categories of wells in EPA's UIC program. If authorized or mandated to regulate hydraulic fracturing broadly under SDWA, EPA might regulate hydraulic fracturing as a Class II activity, which would parallel its approach for regulating the injection of diesel for fracturing purposes. However, it is possible that EPA could classify oil and gas production wells that are hydraulically fractured under a different class, or develop an entirely new regulatory structure or subclass of wells.[39]

A Class II well may be used to dispose of brines (salt water) and other fluids associated with oil and gas production or storage, to store natural gas, or to inject fluids for enhanced oil and gas recovery. EPA estimates that some 80% of Class II wells are enhanced recovery (ER) wells. These wells inject brine, water, stream polymers, or carbon dioxide primarily into oil-bearing formations (also called secondary or tertiary recovery). Enhanced recovery wells are separate from, and typically surrounded by, production wells.[40] **Table 2** outlines basic requirements for Class II wells.

---

[39] Regulatory requirements for wells related to oil and gas production (Class II wells) are located at 40 C.F.R. Parts 144 and 146.

[40] EPA historically has differentiated Class II wells from production wells. The agency's UIC website states that "[p]roduction wells bring oil and gas to the surface; the UIC Program did not regulate production wells." U.S. Environmental Protection Agency, Class II Wells—Oil and Gas Related Injection Wells (Class II), "What are the types of Class II wells?," http://water.epa.gov/type/groundwater/uic/class2/index.cfm.

**Table 2. Minimum EPA Regulatory Requirements for Class II Wells**

| Requirement | Explanation |
|---|---|
| Permit Required | Yes, except for existing Enhanced Oil Recovery (EOR) wells authorized by rule |
| Life of Permit | Specific period, may be for life of well |
| Area of Review | New wells—¼ mile fixed radius or radius of endangerment |
| Mechanical Integrity Test (MIT) Required | Internal MIT: prior to operation, and pressure test or alternative at least once every five years for internal well integrity. External MIT: cement records may be used in lieu of logs. |
| Other Tests | Annual fluid chemistry and other tests as needed/required by permit |
| Monitoring | Injection pressure, flow rate and cumulative volume, observed weekly for disposal and monthly for enhanced recovery |
| Reporting | Annual |

**Source:** U.S. Environmental Protection Agency, Technical Program Overview: Underground Injection Control Regulations, EPA 816-R-02-025, July 2001, p. 11, 67, and Appendix E.

## State Primacy for UIC Program Administration

Thirty-three states have assumed primacy for the UIC program, EPA has lead implementation authority in 10 states, and authority is shared in the remaining states. EPA directly implements the entire UIC program in several oil and gas producing states, including Kentucky, Michigan, New York, Pennsylvania, and Virginia.[41] **Figure 2** identifies state primacy status for the UIC program.

As noted, for oil and gas related injection operations, under Section 1425, a state may be delegated primary enforcement authority without meeting EPA regulatory requirements for state UIC programs promulgated under Section 1421, provided the state demonstrates that it has an effective program that prevents underground injection that endangers drinking water sources. EPA has issued guidance for approval of state programs under Section 1425.[42] If directed by Congress to regulate hydraulic fracturing as underground injection, this regulatory approach could give states significant flexibility and thus might reduce potential regulatory costs, redundancy, and other possible impacts to the industry and the states.[43]

---

[41] To receive primacy, a state, territory, or Indian tribe must demonstrate to EPA that its UIC program is at least as stringent as the federal standards; the state, territory, or tribal UIC requirements may be more stringent than the federal requirements. For Class II wells, states must demonstrate that their programs are effective in preventing endangerment of underground sources of drinking water (USDWs). Requirements for state UIC programs are established in 40 C.F.R. §§144-147.

[42] U.S. Environmental Protection Agency, *Guidance for State Submissions under Section 1425 of the Safe Drinking Water Act*, Ground Water Program Guidance #19, p. 20, http://www.epa.gov/safewater/uic/pdfs/guidance/guide_uic_guidance-19_primacy_app.pdf.

[43] The House report for the 1980 Safe Drinking Water Act Amendments, H.R. 8117, which established Section 1425, states that "So long as the statutory requirements are met, the states are not obligated to show that their programs mirror either procedurally or substantively the Administrator's regulations." H. Report to accompany H.R. 8117, No. 96-1348, September 19, 1980, p. 5.

**Figure 2. Primacy Status for EPA's UIC Program**

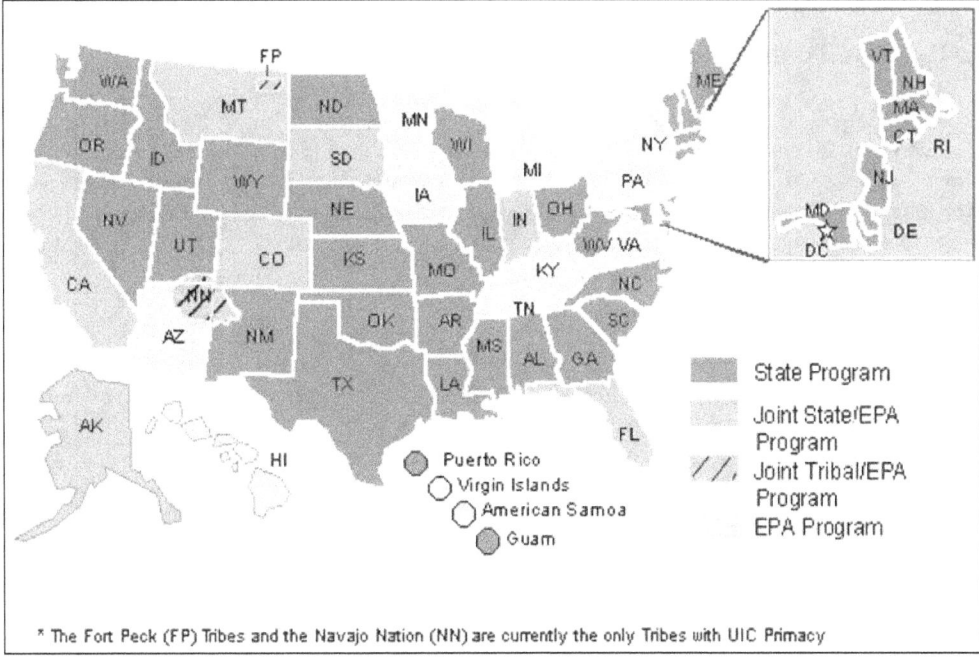

* The Fort Peck (FP) Tribes and the Navajo Nation (NN) are currently the only Tribes with UIC Primacy

**Source:** U.S. Environmental Protection Agency, available at http://www.epa.gov/safewater/uic/primacy.html.

Most oil and gas producing states exercise primary enforcement authority for injection wells associated with oil and gas production (Class II wells) under SDWA Section 1425. Among these states, Alaska, California, Colorado, Indiana, Montana, and South Dakota have received primacy only for Class II wells, while EPA administers the remainder of the UIC program (Class I, III IV, and V wells) for these states. **Table 3** lists states that regulate Class II wells under Section 1425.

**Table 3. States Regulating Oil and Gas (Class II) UIC Wells
Under SDWA Section 1425**

| | | |
|---|---|---|
| Alabama | Louisiana | Oklahoma |
| Alaska | Mississippi | Oregon |
| Arkansas | Missouri | South Dakota |
| California | Montana | Texas |
| Colorado | Nebraska | Utah |
| Illinois | New Mexico | West Virginia |
| Indiana | North Dakota | Wyoming |
| Kansas | Ohio | |

**Source:** Adapted from information provided by U.S. Environmental Protection Agency.

**Note:** With primacy granted under Section 1425, states regulate Class II wells using their own program requirements rather than following EPA regulations, providing significant regulatory flexibility to the states. EPA notes that state requirements "can be, and often are, more stringent than minimum federal standards." Underground Injection Control 101P Permitting Guidance for Hydraulic Fracturing Using Diesel Fuels, Technical Webinars, Mary 9-16, 2011.

# The Debate over Regulation of Hydraulic Fracturing Under the SDWA

From the date of the enactment of the SDWA in 1974 until the late 1990s, hydraulic fracturing was not regulated under the act by EPA or the states tasked with administration of the SDWA. However, in the last 15 years a number of developments have called into question the extent to which hydraulic fracturing should be considered an "underground injection" to be regulated under the SDWA. One trigger for this debate was a challenge to the Alabama UIC program brought by the Legal Environmental Assistance Foundation (LEAF).

## The LEAF Challenge to the Alabama UIC Program and EPA's Interpretation of the SDWA

In 1994, LEAF petitioned EPA to initiate proceedings to have the agency withdraw its approval of the Alabama UIC program because the program did not regulate hydraulic fracturing operations in the state associated with production of methane gas from coalbed formations.[44] The state of Alabama had previously been authorized by EPA to administer a UIC program pursuant to the terms of the SDWA.[45] EPA denied the LEAF petition in 1995 based on a finding that hydraulic fracturing did not fall within the definition of "underground injection" as the term was used in the SDWA and the EPA regulations promulgated under that act.[46] According to EPA, that term applied only to wells whose "principal function" was the placement of fluids underground.[47] LEAF challenged EPA's denial of its petition in the U.S. Court of Appeals for the Eleventh Circuit, arguing that EPA's interpretation of the terms in question was inconsistent with the language of the SDWA.[48]

The court rejected EPA's claim that the language of the SDWA allowed it to regulate only those wells whose "principal function" was the injection of fluids into the ground. EPA based this claim on what it perceived as "ambiguity" in the SDWA regarding the definition of "underground injection" as well as a perceived congressional intent to exclude wells with primarily non-injection functions.[49] The court held that there was no ambiguity in the SDWA's definition of "underground injection" as "the subsurface emplacement of fluids by well injection," noting that the words have a clear meaning and that:

> The process of hydraulic fracturing obviously falls within this definition, as it involves the subsurface emplacement of fluids by forcing them into cracks in the ground through a well. Nothing in the statutory definition suggests that EPA has the authority to exclude from the reach of the regulations an activity (i.e. hydraulic fracturing) which unquestionably falls within the plain meaning of the definition, on the basis that the well that is used to achieve

---

[44] Legal Environmental Assistance Foundation, Inc. v. U.S. Environmental Protection Agency, 118F.3d 1467, 1471 (11th Cir. 1997) ("*LEAF I*").

[45] *Id.* at 1470.

[46] *Id.* at 1471.

[47] *Id.*

[48] *Id.* at 1472.

[49] *Id.* at 1473-74.

that activity is also used—even primarily used—for another activity (i.e. methane gas production) that does not constitute underground injection.[50]

The court therefore remanded the decision to EPA for reconsideration of LEAF's petition for withdrawal of Alabama's UIC program approval.[51]

## Alabama's Regulation of Hydraulic Fracturing in CBM Production

Consideration of Alabama's UIC program after the *LEAF I* decision was issued in 1997 is a helpful case study. It is useful in assessing exactly how EPA authorized a state to regulate hydraulic fracturing under the SDWA "Class" well system, understanding the regulatory options available to EPA and the states authorized to enforce SDWA programs, and evaluating the industry impact resulting from the requirement that hydraulic fracturing be regulated under a UIC program.

Following the *LEAF I* decision and EPA's initiation of proceedings to withdraw its approval of Alabama's Class II UIC program, in 1999 Alabama submitted a revised UIC program to EPA.[52] The revised UIC program sought approval under Section 1425 of the SDWA rather than Section 1422(b). As discussed above, Section 1425 differs from Section 1422(b) in that approval under Section 1425 is based on a showing by the state that the program meets the generic requirements found in Section 1421(b)(1)(A)-(D) of the SDWA and that the program "represents an effective program (including adequate recordkeeping and reporting) to prevent underground injection which endangers drinking water sources."[53] In contrast, approval of a state program under Section 1422(b) requires a showing that the state's program satisfies the requirements of the UIC regulations promulgated by EPA.[54] In its decision on the challenge to EPA's approval of Alabama's revised UIC program, the U.S. Court of Appeals for the Eleventh Circuit observed that "the practical difference between the two statutory methods for approval is that the requirements for those programs covered under § 1425 are more flexible than the requirements for those programs covered under § 1422(b)."[55]

EPA approved Alabama's revised UIC program under Section 1425 in 2000.[56] LEAF appealed EPA's decision to the U.S. Court of Appeals for the Eleventh Circuit. LEAF made three arguments. First, LEAF claimed that EPA should not have approved state regulation of hydraulic fracturing under Section 1425 because it does not "relate to ... underground injection for the secondary or tertiary recovery of oil or natural gas," one of the requirements for approval under Section 1425.[57] The court rejected this argument, finding that the phrase "relates to" was broad

---

[50] *Id.* at 1474-75.

[51] *Id.* at 1478.

[52] *See* 64 Fed. Reg. 56986 (October 22, 1999).

[53] 42 U.S.C. §300h-4(a).

[54] *Id.* at §300h-1(b)(1)(A).

[55] Legal Environmental Assistance Foundation, Inc. v. U.S. Environmental Protection Agency, 276 F.3d 1253, 1257 (11th Cir. 2001) (*LEAF II*).

[56] 65 Fed. Reg. 2889 (October 2000).

[57] *Id.* at 1256.

and ambiguous enough to include regulation of hydraulic fracturing as being related to secondary or tertiary recovery of oil or natural gas.[58]

Second, LEAF challenged the Alabama program's regulation of hydraulic fracturing as "Class II-like" wells not subject to the same regulatory requirements as Class II well.[59] The court agreed with LEAF on this point, noting that in its decision in *LEAF I*, it had held that methane gas production wells used for hydraulic fracturing are "wells" within the meaning of the statute.[60] As a result, the court found that wells used for hydraulic fracturing must fall under one of the five classes set forth in the EPA regulations at 40 C.F.R. Section 144.6.[61] Specifically, the court found that the injection of hydraulic fracturing fluids for recovery of coalbed methane "fit squarely within the definition of Class II wells," and as a result the court remanded the matter to EPA for a determination of whether Alabama's updated UIC program complied with the requirements for Class II wells.[62]

Finally, LEAF alleged that even if Alabama's revised UIC program was eligible for approval under Section 1425 of the SDWA, EPA's decision to approve it was "arbitrary and capricious" and therefore a violation of the Administrative Procedure Act.[63] The court rejected this argument.[64]

Among other provisions added in response to the Eleventh Circuit's decisions, the Alabama regulations prohibited fracturing "in a manner that would allow the movement of fluid containing any contaminant into a USDW, if the presence of the contaminant may (a) cause a violation of any applicable primary drinking water standard; or (b) otherwise adversely affect the health of persons."[65] The state regulations further required state approvals (but not permits) prior to individual fracturing jobs. Specifically, well operators were required to certify in writing, with supporting evidence, that a proposed hydraulic fracturing operation would not occur in a USDW, or that the mixture of fracturing fluids would meet EPA drinking water standards. Regulations also prohibited fracturing at depths shallower than 399 feet (most drinking water wells rely on shallow aquifers) and prohibited the use of diesel oil or fuel in any fracturing fluid mixture. The requirements regarding minimum depths and the diesel ban remain in place, but the rules no longer require that injection fluids meet drinking water standards. Instead, "each coal bed shall be hydraulically fractured so as not to cause irreparable damage to the coalbed methane (CBM) well, or to adversely impact any fresh water supply well or any fresh water resources."[66]

With hydraulic fracturing regulations in place, CBM development in Alabama continued. In 2009, a member of the State Oil and Gas Board of Alabama noted, "since Alabama adopted its

---

[58] *Id.* at 1259-61.

[59] *Id.* at 1256.

[60] *Id.* at 1262.

[61] *Id.* at 1263.

[62] *Id.* at 1263-64.

[63] *Id.* at 1256 (referring to 5 U.S.C. §706(2)(A)).

[64] *Id.* at 1265.

[65] Ala. Admin. Code, r. 400-3-8-.03(4), (2002). Responding to EPAct 2005 (see below), the state made some revisions to its regulations for hydraulic fracturing of coal beds in 2007. Ala. Admin. Code r. 400-3-8-.03(1).

[66] Ala. Admin. Code r. 400-3-8-.03(1).

hydraulic fracturing regulations, coalbed operators have submitted thousands of hydraulic fracturing proposals and engaged in thousands of hydraulic fracturing operations."[67]

The number of CBM well permits increased in the years following the adoption of revised regulations.[68] However, it is not clear whether, or by how much, the number of wells, the production costs, or the time required by operators may have been different without the revisions.[69] One of the requirements of the Alabama regulations in response to *LEAF I* was that fracturing fluids had to meet tap water standards where fracturing would occur within an underground source of drinking water. To ensure compliance, operators purchased water from municipal water supplies that were in compliance with federal drinking water standards to use for fracturing wells. Industry representatives have noted that if this approach were adopted for hydraulic fracturing nationwide, it would not only raise costs, but potentially put companies in competition with communities for drinking water supplies.

Some concern has been expressed that if Congress passed legislation requiring federal regulation of hydraulic fracturing broadly,[70] a separate permit might be required each time a well is hydraulically fractured, thus repeatedly disrupting oil and gas production activities. In Alabama, in response to *LEAF I*, the state did not require a permit for each fracturing operation, but rather had operators give notice and receive approval before fracturing. To further facilitate approvals for hydraulic fracturing, service companies identified to the state chemicals contained in various fracturing fluid mixtures that met the regulatory requirement that the mixtures not exceed federal drinking water standards. A well operator then could select from a list of pre-approved hydraulic fracturing fluids and provide the product name to the state, rather than have to submit separate analyses. Alabama regulations apply this approach where fracturing would occur within an underground source of drinking water.

## EPA's 2004 Review of Hydraulic Fracturing for CBM Production

In response to the *LEAF I* decision, citizen reports of water well contamination attributed to hydraulic fracturing of coal beds, and the rapid growth in CBM development, EPA undertook a study to evaluate the environmental risks to underground sources of drinking water from hydraulic fracturing practices associated with CBM production. EPA issued a draft report in August 2002.[71] The draft report identified water quality and quantity problems that individuals had attributed to hydraulic fracturing of coal beds in Alabama, New Mexico, Colorado, Wyoming, Montana, Virginia, and West Virginia.[72] Based on the preliminary results of the study, EPA tentatively concluded that the potential threats to public health posed by hydraulic fracturing of coalbed methane wells appeared to be small and did not justify additional study or regulation.

---

[67] S. Marvin Rogers, State Oil and Gas Board of Alabama and Chairman, IOGCC Legal and Regulatory Affairs Committee, *History of Litigation Concerning Hydraulic Fracturing to Produce Coalbed Methane*, January 2009, p. 5.

[68] Ala. Admin. Code r. 400-3-8-.03(6)(a), 2002. To mitigate its increased administrative costs associated with implementation of the added regulations, operators pay a fee of $175 for each coalbed group fractured.

[69] A representative of the Alabama Coalbed Methane Association noted that the costs of hydraulic fracturing are very site specific and vary with operators as well as geology.

[70] Currently, EPA has authority to regulate only the use of diesel fuel in fracturing operations.

[71] U.S. Environmental Protection Agency. Draft Evaluation of Impacts to Underground Sources of Drinking Water by Hydraulic Fracturing of Coalbed Methane Reservoirs. EOA 816-D-02-006, August 2002.

[72] *Id.*, p. 6-20-6-21.

---

EPA also reviewed whether direct injection of fracturing fluids into underground sources of drinking water posed any threat. EPA reviewed 11 major coalbed methane formations to determine whether coal seams lay within USDWs. EPA determined that 10 of the 11 producing coal basins "definitely or likely lie entirely or partially within USDWs."

In January 2003, the EPA's National Drinking Water Advisory Council submitted to the EPA Administrator a report on hydraulic fracturing, underground injection control, and coalbed methane production and its impacts on water quality and water resources. The Council noted concerns regarding (1) the lack of resources to implement the UIC program, (2) the use of diesel fuel and potentially toxic additives in the hydraulic fracturing process, (3) the potential impact of coalbed methane development on local underground water resources and the quality of surface waters, and (4) the maintenance of EPA regulatory authority within the UIC program. The Council recommended that EPA:

- work through regulatory or voluntary means to eliminate the use of diesel fuel and related additives in fracturing fluids that are injected into formations containing sources of drinking water;

- continue to study the extent and nature of public health and environmental problems that could occur as a result of hydraulic fracturing for coalbed methane production; and

- defend EPA's existing authority and discretion to implement the UIC program in a manner that advances the protection of groundwater resources from contamination.[73]

In 2004, EPA issued a final version of the 2002 draft report, based primarily on an assessment of the available literature and extensive interviews, and concluded that the injection of hydraulic fracturing fluids into CBM wells posed little threat to underground sources of drinking water and required no further study. However, EPA found that very little documented research had been done on the environmental impacts of injecting fracturing fluids.[74] Additionally, EPA had discussed the use of diesel fuel in fracturing fluids in the 2002 draft report, and concluded in the final report that "The use of diesel fuel in fracturing fluids poses the greatest potential threat to USDWs because the BTEX constituents in diesel fuel exceed the MCL [maximum contaminant level] at the point-of-injection."[75]

EPA also noted that estimating the concentration of diesel fuel components and other fracturing fluids beyond the point of injection was beyond the scope of its study.[76] Moreover, the EPA study focused specifically on CBM wells and did not review the use of hydraulic fracturing in other geologic formations, such as the Marcellus and Barnett shales and tight gas sand formations.

---

[73] National Drinking Water Advisory Council. Report on Hydraulic Fracturing and Underground Injection Control and Coalbed Methane by the National Drinking Water Advisory Council Resulting from a Conference Call Meeting Held December 12, 2002. Washington DC.

[74] U.S. Environmental Protection Agency, *Evaluation of Impacts to Underground Sources of Drinking Water by Hydraulic Fracturing of Coalbed Methane Reservoir*s, Final Report, EPA-816-04-003, Washington, D.C., June 2004, p. 4-1.

[75] *Evaluation of Impacts to USDWs by Hydraulic Fracturing of Coalbed Methane Reservoirs*, Final Report, p. 4-19.

[76] *Id.* p. 4-12.

To address concerns about the use of diesel fuel in hydraulic fracturing fluids, EPA entered into an agreement with three companies that provided roughly 95% of hydraulic fracturing services (BJ Services, Halliburton Energy Services, and Schlumberger Technology Corporation). Under this agreement, the firms agreed to remove diesel fuel from CBM fluids injected directly into drinking water sources if cost-effective alternatives were available.[77]

# EPAct 2005: A Legislative Exemption for Hydraulic Fracturing

The decision by the U.S. Court of Appeals for the Eleventh Circuit in *LEAF I* highlighted a debate over whether the SDWA as it read at the time required EPA to regulate hydraulic fracturing. Although the Eleventh Circuit's decision applied only to hydraulic fracturing for coalbed methane production in Alabama, the court's reasoning—in particular, its finding that hydraulic fracturing "unquestionably falls within the plain meaning of the definition [of underground injection]"[78]—raised the issue of whether EPA could be required to regulate hydraulic fracturing under the SDWA.

Before this question was resolved through agency action or litigation, Congress passed an amendment to the SDWA as a part of EPAct 2005 (P.L. 109-58) that addressed this issue. Section 322 of EPAct 2005 amended the definition of "underground injection" in the SDWA as follows:

> The term "underground injection"—(A) means the subsurface emplacement of fluids by well injection; and (B) excludes—(i) the underground injection of natural gas for purposes of storage; and (ii) the underground injection of fluids or propping agents (other than diesel fuels) pursuant to hydraulic fracturing operations related to oil, gas, or geothermal production activities.

This amendment clarified that the UIC requirements found in the SDWA do not apply to hydraulic fracturing, although the exclusion does not extend to the use of diesel fuel in hydraulic fracturing operations. This amended language is the definition of "underground injection" found in the SDWA as of the date of this report.

In 2010, EPA posted on its website a requirement that any service company that performs hydraulic fracturing using diesel fuel must receive prior authorization from the relevant UIC authority (state or EPA). EPA also determined that injection wells receiving diesel fuel are Class II wells for purposes of the UIC program. This determination was made without the development of guidance or regulations (and without notice and comment as would normally be done in an agency rulemaking proceeding).[79] EPA's approach created uncertainty in the industry and among oil and gas and UIC regulators. Currently, EPA is developing UIC Class II permitting guidance to assist states in regulating hydraulic fracturing operations that use diesel fuels. EPA notes that the

---

[77] *Memorandum of Agreement Between the United States Environmental Protection Agency and BJ Services Company, Halliburton Energy Services, Inc., and Schlumberger Technology Corporation,* December 12, 2003.

[78] *LEAF I,* 118 F.3d at 1475.

[79] http://water.epa.gov/type/groundwater/uic/class2/hydraulicfracturing/wells_hydroreg.cfm#safehyfr. The website notes that "[a]ny service company that performs hydraulic fracturing using diesel fuel must receive prior authorization from the UIC program," and that "[i]njection wells receiving diesel fuel as a hydraulic fracturing additive will be considered Class II wells by the UIC program."

guidance will be based on existing regulations and will contain recommendations for permit writers to consider when writing permits for production wells where diesel fuel is used as a fracturing fluid.[80]

## Proposed Legislation in the 112ᵗʰ Congress

In March 2011, the Fracturing Responsibility and Awareness of Chemicals Act of 2011 (FRAC Act), H.R. 1084 and S. 587, was introduced in the Senate and the House of Representatives.[81] The bills have some minor language differences, but are substantially similar. (They also are similar to bills introduced in the past Congress.) Each contains two amendments to the SDWA—one that would amend the definition of underground injection to include hydraulic fracturing, and another that would create a new disclosure requirement for the chemicals used in hydraulic fracturing.

### FRAC Act

H.R. 1084 provides that the definition of "underground injection" that was amended in 2005 to exclude most hydraulic fracturing would be amended once again to include "the underground injection of fluids or propping agents pursuant to hydraulic fracturing operations related to oil, gas or geothermal production activities," excluding injection of natural gas for subsurface storage.[82] This would not only repeal the amended definition of "underground injection" that was enacted as part of EPAct 2005 which excluded hydraulic fracturing, but would essentially codify the court's decision in *LEAF I* and clear up any ambiguity regarding regulation of hydraulic fracturing under the SDWA.

The second amendment to the SDWA in the FRAC Act would create a new hydraulic fracturing disclosure requirement.[83] H.R. 1084 would create a new statutory obligation requiring anyone conducting hydraulic fracturing to:

> disclose to the State (or the [EPA] if the [EPA] has primary enforcement responsibility in the State)—(I) prior to the commencement of any hydraulic fracturing operations at any lease area of portion thereof, a list of chemicals intended for use in any underground injection during such operations, including identification of the chemical constituents of mixtures, Chemical Abstracts Service numbers for each chemical and constituent, material safety data sheets when available, and the anticipated volume of each chemical; and (II) not later than 30 days after the end of any hydraulic fracturing operations the list of chemicals used in each underground injection during such operations, including identification of the chemical constituents of mixtures, Chemical Abstracts Service numbers for each chemical and constituent, material safety data sheets when available, and the volume of each chemical used.[84]

---

[80] U.S. Environmental Protection Agency, *Underground Injection Control Guidance for Permitting Oil and Natural Gas Hydraulic Fracturing Activities Using Diesel Fuels*, http://water.epa.gov/type/groundwater/uic/class2/hydraulicfracturing/wells_hydroout.cfm.

[81] H.R. 1084, S. 587.

[82] H.R. 1084, at §2(a). S. 587 is similar but does not include geothermal production activities.

[83] For a detailed review of state and federal chemical disclosure developments, see CRS Report R42461, *Hydraulic Fracturing Chemical Disclosure Requirements*, by Brandon J. Murrill and Adam Vann.

[84] *Id.* at §2(b).

The bill would also require that the state or EPA "make the disclosure of chemical constituents ... available to the public, including by posting the information on an appropriate Internet Web site," and the bill clarifies that the disclosure requirements "do not authorize the State (or the [EPA]) to require the public disclosure of proprietary information."[85] In other words, the disclosure requirements address only the chemicals used, not the manner of their use or the amounts or ratios in which they are used. This language attempts to protect proprietary business information, that is, "secret" formulas or practices that drilling companies may feel they should not be required to disclose to their competitors. Some state oil and gas production statutes and regulations extend similar protections for proprietary business information, while still requiring disclosure to regulators of the chemical constituents being used in hydraulic fracturing.[86]

Furthermore, the FRAC Act would require operators to disclose proprietary chemical information to treating medical professionals in cases of medical emergencies.[87] Although most state oil and gas rules do not require disclosure of proprietary chemical information to medical professionals, such disclosure broadly parallels federal requirements under the Occupational Safety and Health Act (OSHAct).[88] Nonetheless, the OSHAct requirements were not designed for environmental

---

[85] *Id.*

[86] In 2008, for example, the Colorado Oil and Gas Conservation Commission promulgated regulations requiring operators to maintain inventories of chemicals stored onsite for use downhole, and to provide a list of the chemicals of "trade secret chemical products" to commission officials upon request. Operators are also required to disclose chemical information to treating medical professionals. (2 Colo. Code Regs. §404-1:205).

Wyoming is another state that did not wait for the federal government to adopt disclosure requirements for persons engaged in hydraulic fracturing. On September 15, 2010, the Wyoming Oil and Gas Conservation Commission (WOGCC) promulgated its own set of hydraulic fracturing disclosure requirements. In accordance with these regulations, drilling operators are required to:

• identify all water supply wells within one-quarter mile of the drilling activity as well as the depth from which water is being appropriated (Wyo. Rules and Regs. Oil Gen §3-8);

• provide stimulation fluid information to the WOGCC on its Application for Permit to Drill, as part of a comprehensive drilling/completion/recompletion plan, or on a separate notice (Wyo. Rules and Regs. Oil Gen §3-45(a));

• provide geological names, geological description and depth of the formation into which well stimulation fluids are to be injected (Wyo. Rules and Regs. Oil Gen §3-45(c));

• provide to an WOGCC Supervisor, for each stage of the well stimulation program, the chemical additives, compounds and concentrations or rates proposed to be mixed and injected, including (i) stimulation fluid identified by additive type; (ii) the chemical compound name and Chemical Abstracts Service (CAS) number of any constituents; and (iii) the proposed rate or concentration for each additive. The WOGCC Supervisor is also authorized to request additional information as deemed appropriate (Wyo. Rules and Regs. Oil Gen §3-45(d)).

• provide a detailed description of the proposed well stimulation design, which shall include (i) the anticipated surface treating pressure range; (ii) the maximum injection treating pressure; and (iii) the estimated or calculated fracture length and fracture height.

The regulations prohibit the underground injection of "volatile organic compounds, such as benzene, toluene, ethylbenzene and xylene, also known as BTEX compounds or any petroleum distillates, into groundwater." (Wyo. Rules and Regs. Oil Gen. §3-45(g)). The regulations do state that confidentiality protection will be provided for "trade secrets, privileged information and confidential commercial, financial, geological or geophysical data furnished by or obtained from any person." (Wyo. Rules and Regs. Oil Gen. §3-45(f)). There also are logging requirements applicable to post-well stimulation (Wyo. Rules and Regs. Oil Gen. §3-45(h)).

[87] H.R. 1084, §2(b).

[88] The Occupational Safety and Health Administration has promulgated a set of regulations under Occupational Safety and Health Act (OSHAct), referred to as the Hazard Communication Standard (29 C.F.R. §1910.1200). Additionally, OSHAct regulations require operators to maintain Material Safety Data Sheets (MSDS) for hazardous chemicals at the job site. The federal Emergency Planning and Community Right to Know Act (EPCRA) requires that facility owners submit an MSDS for each hazardous chemical present that exceeds an EPA-determined threshold level, or a list of such (continued...)

---

investigation purposes and have been criticized as deficient. Calls for disclosure of hydraulic fracturing chemicals have increased as homeowners and others express concern about the potential presence of unknown chemicals in tainted well water near oil and gas operations.

## FRESH Act

Introduced in March 2012, the Fracturing Regulations are Effective in State Hands Act (FRESH Act), S. 2248 and H.R. 4322, would specify that a state has sole authority to regulate hydraulic fracturing on federal lands within the boundaries of the state. This legislation comes after the President announced in his 2012 State of the Union address that he would require "all companies that drill for gas on public lands to disclose the chemicals they use." A draft of the Bureau of Land Management (BLM) proposed rule would require public disclosure of chemicals used in hydraulic fracturing on BLM managed lands.[89] The draft also would require operators to submit to BLM detailed information on their proposed well stimulation plans (such as expected volume of fluid to be used, injection pressures, estimated fracture length, and plans for treating flowback). Updated information would be required after hydraulic fracturing operations.[90]

# Potential Implications of Hydraulic Fracturing Regulation Under the SDWA

The full regulation of hydraulic fracturing under the SDWA (i.e., beyond injections involving diesel) potentially could have significant, but currently unknowable, environmental benefits as well as impacts on oil and natural gas producers and state and federal regulators. Resulting groundwater protection, public health, and economic benefits would likely be experienced most significantly in states that may currently have relatively weak groundwater protection requirements (such as substandard cementing and casing requirements, or permitting injection of unknown chemicals directly into USDWs). However, the specific regulation of the underground injection of fluids for hydraulic fracturing purposes would not address surface management of chemicals, drilling wastes, or treatment and disposal of produced water. If such surface activities are determined to be the sources of most water contamination incidents associated with unconventional oil and gas development, federal regulation of hydraulic fracturing under the SDWA may have limited environmental and public health benefits. Benefits of federal regulation could be significant if various states have oil and gas regulations that are not adequate to address the new drilling and production methods applied to unconventional oil and gas resources.

Regulations requiring chemical disclosure could also be beneficial. The lack of information regarding chemicals used in hydraulic fracturing has made investigations of contamination

---

(...continued)

chemicals, to the local emergency planning committee (LEPC), the state emergency response commission, and the local fire department. For non-proprietary information, EPCRA generally requires a LEPC to provide an MSDS to a member of the public on request.

[89] On federal lands, the Bureau of Land Management, within the Department of the Interior, administers leasing and coordinates planning and permitting with other federal agencies, as appropriate.

[90] Mike Soraghan, Hydraulic Fracturing: BLM Proposes More Disclosure than Most States, Greenwire, February 6, 2012. For further discussion, see CRS Report R42461, *Hydraulic Fracturing Chemical Disclosure Requirements*, by Brandon J. Murrill and Adam Vann.

difficult as well owners and state regulators typically do not know which chemicals to test for to determine whether a fracturing fluid has migrated into a water source.[91]

If the SDWA were amended to authorize (but not mandate) EPA to regulate hydraulic fracturing, EPA likely would need to undertake further study to assess the potential risks of hydraulic fracturing to underground sources of drinking water. (The agency currently is conducting such studies, as discussed below.) Subsequently, EPA might determine the need for, and potential scope of, any new regulations, and decide whether to adapt the existing regulatory framework or to develop a new approach under the UIC program. The rulemaking process typically takes several years. A 2009 presentation by EPA's Region 8 UIC program explained that, if legislative change occurs:

> additional study may take place, regulations may be written by EPA, some combination of these may happen, [and] there may be a phased-in approach. If regulations are developed, they typically include: establishing a regulation development workgroup which can include the public; a proposed regulation, including opportunity for public comment (and one or more hearings if needed); a final regulation, including opportunity for judicial appeals; and an effective date for the regulation.[92]

One implication of regulating hydraulic fracturing under SDWA relates to the SDWA's citizen suit provisions. As noted, Section 1449 provides for citizen civil actions against any person or agency allegedly in violation of provisions of SDWA, or against the EPA Administrator for alleged failure to perform any action or duty that is not discretionary.[93] This provision could represent an expansion in the ability of citizens to challenge state administration of oil and gas programs related to hydraulic fracturing and drinking water, were the hydraulic fracturing exemption provision to be repealed.

As discussed, the SDWA currently includes two options for approving state UIC programs related to oil and gas recovery.[94] Under the less restrictive requirements of Section 1425, EPA may be able to implement new requirements primarily through guidance and review and approval of state programs revised to address hydraulic fracturing. EPA used this approach when ordered to require Alabama to regulate hydraulic fracturing of coal beds, and a federal district court approved this approach.

If EPA decided to allow states to regulate hydraulic fracturing under Section 1425, the agency also might be expected to write new hydraulic fracturing regulations under Section 1421 in order to implement the program directly for states that do not have primacy for Class II wells and for states that exercise primacy under Section 1422. Regardless of regulatory approach, new requirements would likely require substantially more resources for UIC program administration and enforcement by the states and EPA.

---

[91] The GWPC and Interstate Oil and Gas Compact Commission established a public registry where companies may voluntarily identify chemicals used in hydraulic fracturing in individual wells (http://www.fracfocus.org).

[92] U.S. Environmental Protection Agency, Region 8, *Hydraulic Fracturing*, Presentation, Underground Injection Control Program Meeting, Glenwood Springs, Colorado, August 8, 2009.

[93] §1449; 42 U.S.C. 300j-8.

[94] In the case concerning Alabama, the Eleventh Circuit Court of Appeals ruled that "EPA's decision to subject hydraulic fracturing to approval under § 1425 rests upon a permissible construction of the Safe Drinking Water Act." *Legal Environmental Assistance Fund v. Environmental Protection Agency, State Oil and Gas Board of Alabama*, 276 F.3d 1253 (11th Cir. 2001).

The possible impacts of enacting legislation directing EPA to regulate hydraulic fracturing could vary for different oil and gas production operations. The SDWA directs EPA, when developing UIC regulations, to take into consideration "varying geologic, hydrological, or historical conditions in different States and in different areas within a State."[95] Consequently, if EPA were to regulate hydraulic fracturing under the SDWA, the agency conceivably could establish different requirements to address such differences among states or regions. If practical and applicable, EPA might find this statutory flexibility helpful, as the USDW contamination risks of hydraulic fracturing could vary widely among different formations and settings. For example, fracturing a coal bed that may qualify as a USDW poses very different groundwater contamination risks than fracturing a shale formation that is widely separated from any USDW.[96] Thus, the possible application and impact of federal regulations might vary significantly in different formations, and the impacts and potential environmental benefits would likely be greatest in such coal beds or other formations occurring in or near USDWs.[97] However, EPA might broadly apply other requirements, such as those related to well construction and cementing, and mechanical integrity testing, to protect USDWs through which wells may pass, among other purposes.

For the oil and gas industry, regulation of hydraulic fracturing under the UIC program could have a range of impacts. In some states, oil and gas operations are subject to regulation by a state oil and gas agency or commission as well as an environmental or public health agency. Industry representatives have expressed concern over the potential for some duplication of requirements from state oil and gas regulators and environmental regulators. Delays in issuing permits, and commensurate delays in well stimulation and gas marketing are among the concerns. The citizen suit provision of the SDWA also may be an issue. One analysis attempting to measure the economic and energy effects of potential regulation noted that:

> Experience suggests that there will be a reduction in the number of wells completed each year due to increased regulation and its impact on the additional time needed to file permits, push-back of drilling schedules due to higher costs, increased chance of litigation, injunction or other delay tactics used by opposing groups and availability of fracturing monitoring services.[98]

Several studies have attempted to estimate the potential economic and energy supply impact of regulating hydraulic fracturing under the federal UIC program. A 2009 study prepared by a consultant for DOE estimated the costs associated with "a stringent set of potential federal requirements" including (1) obtaining a permit, (2) conducting an area of review assessment, (3) performing in-situ stress analysis, (4) conducting three-dimension fracture simulation, (5) monitoring, (6) mapping fractures, or conducting other post-fracture analysis, (7) for some wells (perhaps 10%), performing state-of-the-art down-hole fracture imaging, and (8) additional cement to ensure isolation of the target zone before fracturing.[99] Based on these assumed elements of a

---

[95] §1421(b)(3)(A); 42 U.S.C. 300h(b)(3)(A).

[96] Because coal beds frequently are sources of drinking water, the Alabama State Oil and Gas Board requires well operators to certify that a proposed hydraulic fracturing operation would not occur in a USDW, or that the mixture of fracturing fluids would meet EPA drinking water standards. The state regulations also prohibit fracturing at depths shallower than 399 feet, as most drinking water wells rely on shallow aquifers.

[97] U.S. Department of Energy Office of Fossil Energy and National Technology Laboratory, *Modern Shale Gas Development in the United States A Primer*, DE-FG26-04NT15455, April 2009, http://fossil.energy.gov/programs/oilgas/publications/naturalgas_general/Shale_Gas_Primer_2009.pdf.

[98] IHS Global Insight, *Measuring the Economic and Energy Impacts of Proposals to Regulate Hydraulic Fracturing*, Task 1 Report, Prepared for the American Petroleum Institute, Lexington, MA, 2009, p. 7.

[99] Advanced Resources International, Inc., *Potential Economic and Energy Supply Impacts of Proposals to Modify* (continued...)

regulatory program, the study estimated that the compliance costs for regulating hydraulic fracturing for oil and gas development would be $100,505 for new wells receiving hydraulic fracturing treatment.[100]

A stringent regulatory program under Section 1422 arguably could include many of the above requirements. However, it is unknown what EPA might require and unclear what costs would be attributed to federal regulation. Some activities already are used in the industry or required by states (e.g., well cementing across all groundwater zones). EPA UIC staff note that some of the requirements assumed in the study have never been a part of the federal UIC regulations. Other effects that are not easily quantified include the costs associated with waiting periods between fracturing jobs for approvals and other potential disruptions to operations.

The Ground Water Protection Council (GWPC),[101] representing state agencies, has opposed reclassification of hydraulic fracturing as a permitted activity under the UIC programs, stating that (1) a risk has not been identified, and thus, there is no evidence that regulation is necessary; and (2) UIC regulation would divert resources from higher risk activities.[102] The legislatures of major oil and gas producing states, including the states of Alabama, Alaska, Montana, North Dakota, Wyoming, and Texas, passed and sent to Congress resolutions asking Congress not to extend SDWA jurisdiction over hydraulic fracturing activities.

As discussed, the GWPC is recommending the adoption of various best management practices to strengthen protection of water resources when developing oil and gas resources. Industry appears to be adopting some of these management practices independent of regulation. In discussing lessons learned from developing the Barnett shale, industry consultants recently reported, that an "important factor, requiring 3D seismic [imaging], is the avoidance of geo-hazards, such as water-bearing karsts and faults."[103] However, voluntary industry practices can not be enforced, and there is no assurance that they would be widely adopted.

If authorized, EPA regulation of hydraulic fracturing under the SDWA UIC program would not address many public concerns often associated with the development of unconventional oil and gas resources. These concerns involve land surface disturbances associated with the development of roads, well pads, and natural gas gathering pipelines; potential impacts of water withdrawal; treatment and disposal of flowback water to surface waters; air quality impacts; etc. Some of these activities are subject to other federal laws, such as Clean Water Act requirements covering

---

(...continued)

*Federal Environmental Laws Applicable to the U.S. Oil and Gas Exploration and Production Industry*, U.S. Department of Energy, Office of Fossil Energy, January 2009. The authors note that cost estimates are based on a 1999 memorandum prepared for DOE, from Robin Petrusak, ICF Consulting to Nancy Johnson, U.S. Department of Energy, "Documentation of Estimated Potential Cost of Compliance for Toxic Release Inventory (TRI) Reporting and Hydraulic Fracturing," August 19, 1999.

[100] *Id.* p. 25-26.

[101] The GWPC is a national association representing state groundwater and UIC agencies whose mission is to promote protection and conservation of groundwater resources for beneficial uses. The stated purpose of the GWPC is "to promote and ensure the use of best management practices and fair but effective laws regarding comprehensive ground water protection." http://www.gwpc.org/about_us/about_us.htm.

[102] Statement of Scott Kell, for the Ground Water Protection Council, House Committee on Natural Resources, Subcommittee on Energy and Mineral Resources, Oversight Hearing on "Unconventional Fuels, Part I: Shale Gas Potential," June 4, 2009.

[103] Scott Stevens and Vello Kuuskraa, Advanced Resources International, Inc., "Gas Shale-1: Seven Plays Dominate North America Activity," *Oil & Gas Journal*, vol. 107, no. 36 (September 28, 2009), p. 41.

---

the treatment and discharge of produced water into surface waters. The regulatory impacts of state and federal regulatory requirements for treatment and discharge of produced water may be more significant than potential UIC-related requirements.[104] Other impacts related to development of unconventional oil and gas resources are highly visible and may raise more issues than the specific process of deep underground fracturing of oil and gas formations. Some of these issues (particularly certain land-use and mineral resource development issues) are beyond the reach of federal regulation, and thus, are left to state and local governments to address.[105] New York State's Revised Draft Supplemental Generic Environmental Impact Statement is one example of a state taking a comprehensive approach to addressing a broad range of possible environmental impacts that could be associated with Marcellus Shale development. [106]

# State Regulation of Hydraulic Fracturing

While the federal government currently exempts most hydraulic fracturing activity from regulation under the SDWA, the states are free to regulate the practice as they see fit. Although state oil and gas regulatory programs initially focused on managing petroleum reservoirs, efficient production, and addressing mineral rights issues, these programs have become more environmentally focused through the decades. The GWPC and the Interstate Oil and Gas Compact Commission (IOGCC) [107] each report that the major oil and gas producing states now have laws and regulatory requirements in place to protect water resources during oil and natural gas exploration and production activities.

Both the GWPC and the IOGCC oppose federal regulation of hydraulic fracturing, noting that this process is regulated by the states, sometimes specifically, but most often through general oil and gas production regulations, policies, and practices.[108] The IOGCC notes that member states have adopted comprehensive laws and regulations to provide for safe operations and to protect the nation's drinking water sources, and that these states have trained personnel with expertise to effectively regulate oil and gas exploration and production; thus, making the states the best-suited regulators of hydraulic fracturing. The IOGCC further makes the case for keeping responsibility with the states:

---

[104] See *Marcellus Shale Gas  Development Potential and Water Management Issues and Laws* at 15.

[105] For a review of the applicability of various federal environmental laws to oil and gas development, see Amy Mall, Natural Resources Defense Council, "The applicability of federal requirements that protect public health and the environment to oil and gas development," Testimony before the Committee on Oversight and Government Reform, U.S. House of Representatives, October 31, 2007, http://www.nrdc.org/energy/ene_07103101.asp.

[106] New York imposed a temporary moratorium on unconventional gas drilling until the state can update oil and gas regulations to govern development of the Marcellus Shale and other tight shale formations in the state using hydraulic fracturing combined with directional drilling. See New York State Department of Environmental Conservation and Division of Mineral Resources, *Revised Draft Supplemental Generic Environmental Impact Statement on the Oil, Gas and Solution Mining Regulatory Program  Well Permit Issuance for Horizontal Drilling and High-Volume Hydraulic Fracturing to Develop the Marcellus Shale and Other Low-Permeability Gas Reservoirs*, September 2011, http://www.dec.ny.gov/energy/75370.html.

[107] The Interstate Oil and Gas Compact Commission represents the state oil and gas agencies. The commission was established in the 1930s, initially to reduce the waste of oil during exploration and production by developing model statutes and practices to improve the conservation of oil resources.

[108] The GWPC passed a resolution in 2003 encouraging Congress to clarify the definition of underground injection in Part C of SDWA to exclude the practice of hydraulic fracturing. http://www.gwpc.org/advocacy/documents/resolutions/RES-03-5.htm.

---

Hydraulic fracturing is currently, and has been for decades, a common operation used in exploration and production by the oil and gas industry in all gas producing states. Because of the unique position of the states and their collective expertise on matters concerning the oil and gas industry, regulation of hydraulic fracturing should remain the responsibility of the States. The States have as much of a vested interest in the protection of groundwater as the federal government and as such, will continue to regulate the process effectively and efficiently, taking into account the particulars of the geology and hydrology within their boundaries. There is not a "one-size fits all" approach to effective regulation.[109]

The question that has arisen is whether state oil and gas programs effectively address increasing groundwater protection concerns arising with the heightened concentration and broadened geographic extent of unconventional and conventional oil and gas resource development that relies on hydraulic fracturing in combination with deep horizontal drilling.[110] As noted, various states recently have revised their oil and gas exploration and production regulations in response to new types and levels of natural gas production, and specifically to increase protection of water resources.

A related issue concerns the extent to which state oil and gas agencies coordinate adequately with their water pollution control counterparts. Most states have different agencies administering oil and gas programs and environmental programs. State UIC programs often are administered by the environmental agency, while oil and gas exploration and production activities are overseen by separate oil and gas entities. Moreover, with the exception of Alabama, which acted in response to a court ruling, no state has chosen to regulate hydraulic fracturing as part of its EPA-authorized underground injection control program.[111]

## GWPC Review of State Regulations

The Ground Water Protection Council Although states have extensive regimes in place to manage oil and gas development activities, the GWPC also noted that related state groundwater protection regulations, policies, and practices can be uneven. In 2009, the GWPC published a review of state oil and gas regulations designed to protect water resources for the 27 major oil and gas producing states.[112] Based on this review, the GWPC concluded that, in general, state oil and gas regulations are adequately designed to protect water resources. Among the states, requirements to protect water resources covered permitting, well drilling, and construction (e.g., casing, cementing, and test pressure requirements), well closure and abandonment, and waste fluid management.

---

[109] Further policy positions and information can be found at the IOGCC website: http://www.iogcc.org/hydraulic-fracturing.

[110] Hydraulic fracturing is used commonly used for conventional gas production. Wyoming, for example, reported that in 2008, 100% (1,316) of new conventional gas wells were fracture stimulated, many wells with multi-zone stimulations in each well bore, some staged, and some individual fracture stimulations. Source: Wyoming Oil and Gas Conservation Commission. The Commission rules require operators to receive approval prior to hydraulic fracturing treatments. Operators are required to provide detailed information regarding the fracturing process, including the source of water and/or trade name fluids, type of proppants, and estimated pump pressures. After a treatment is complete, the operator must provide fracturing data and production results.

[111] In October 2007, in response to the 2005 Energy Policy Act, Alabama revised its Class II UIC program to once again exclude hydraulic fracturing. The state retains most hydraulic fracturing requirements which it administers under its oil and gas regulatory regime.

[112] Ground Water Protection Council, U.S. Department of Energy, Office of Fossil Energy, National Energy Technology Laboratory, *State Oil and Natural Gas Regulations Designed to Protect Water Resources*, May 2009.

---

While few states explicitly mentioned hydraulic fracturing in their regulations, many had well drilling, construction, completion, and reporting requirements intended to protect ground and surface water resources. For example, 10 major producing states required reporting of chemicals used in well treatments, 25 states required operators to submit well treatment (including fracturing) reports, and 22 states require operators to cement across groundwater zones. State requirements vary greatly, from the specific requirements in Alabama to more general mandates not to harm water resources (e.g., Arizona oil and gas rules require operators to "conduct operations in a manner that prevents oil, gas, salt water, fracturing fluid or any other substance from polluting any surface or subsurface waters"). Colorado's regulations include a well casing program to protect groundwater (and hydrocarbons), require well treatment and fracturing reporting, and require operators to notify landowners at least one week before conducting various operations, including fracturing.[113]

While finding that most states had an extensive array of permitting and operating requirements for oil and gas wells, the GWPC also noted that some states lacked important provisions in their programs. For example, most, but not all, states had well construction requirements that include provisions for cementing above oil or gas producing zones and across groundwater zones. The GWPC made a series of recommendations to strengthen state programs to protect water resources. A sample of findings and recommendations from the GWPC review follows:

- State oil and gas regulations are adequately designed to directly protect water resources through the application of specific program elements such as permitting, well construction, well plugging, and temporary abandonment requirements.

- Some exploration and production (E&P) activities have caused contamination of both surface and groundwater. Past practices related to pit construction, well cementing and operation, and well plugging were not always adequate to prevent migration of contaminants to surface and groundwater.

- Hydraulic fracturing in oil or gas bearing zones that occur in non-exempt USDWs should be either stopped, or restricted to the use of materials that do not pose a risk of endangering groundwater and do not have the potential to cause human health effects (e.g., fresh water, sand, etc.).

- Hydraulic fracturing of deep zones poses little to no risk of groundwater contamination.

- States should review current regulations in several program areas to determine whether they meet an appropriate level of specificity (e.g., use of standard cements, plugging materials, pit liners, siting criteria, and tank construction standards, etc.).

- Comprehensive studies should be undertaken to determine the relative risk to groundwater resources from the practice of shallow hydraulic fracturing. These studies should be used, with current knowledge, to develop a generic set of best management practices (BMPs) for hydraulic fracturing which state agencies

---

[113] More details of state rules are included in the Regulations Reference Document accompanying the GWPC report, *State Oil and Natural Gas Regulations Designed to Protect Water Resources*, http://www.gwpc.org/e-library/ e_library_list.htm.

could use either to develop state specific BMPs or develop additional state regulations.

- Many states split jurisdiction between oil and gas and water quality or pollution control agencies over some aspects of oil and gas regulation including tanks, pits, waste handling and spills. Where split jurisdiction of oil and gas operations exists, formal memoranda of agreement and regulatory implementation plans should be negotiated.[114]

- States should consider requiring companies to submit a list of additives used in formation fracturing and their concentration within the fracture fluid matrix. Further, states that do not currently regulate handling and disposal of fracture fluid additives and constituents recovered during recycling operations should consider the need to develop such regulations.

- A state program review process, conducted by the national nonprofit group, State Review of Oil and Natural Gas Environmental Regulations (STRONGER),[115] should be recognized as an effective tool for assessing the capability of state programs to manage E&P waste and measure program improvement over time.

- Best Management Practices that can be adapted to each state should be developed to manage hydraulic fracturing. STRONGER should evaluate whether to update its mission to include environmental elements of state oil and gas programs beyond the traditional area of E&P waste [to include hydraulic fracturing]. (STRONGER issued hydraulic fracturing guidelines in February 2010, and review teams are using the guidelines to the evaluate oil and gas regulatory programs of states that have volunteered to be reviewed.)[116]

Various states have determined that the expanding development of unconventional oil and gas resources, along with the increased use of hydraulic fracturing and directional drilling, requires more state oversight. Some states are responding by increasing staff resource levels. And in some states, including Colorado, North Dakota, New York, Ohio, Pennsylvania, Texas, West Virginia, and Wyoming, this expansion has prompted a reassessment and revision of oil and gas production regulations and policies.

## UIC Program Resources

The funding and staffing resource implications of including hydraulic fracturing under the UIC program could be significant for regulatory agencies. The scope of the added workload under Class II UIC programs could more than double. Currently, there are approximately 146,800 Class

---

[114] Four states reported to GWPC that agencies other than the oil and gas authority are involved in the permit review process, either by requirement or upon request of the oil and gas agency. In 2008, Colorado revised its oil and gas regulations to allow for greater public participation in permitting and environmental assessment of oil and gas field sites. This expanded participation includes review by other state water protection agencies. GWPC (2009).

[115] STRONGER, Inc., State Review of Oil and Natural Gas Environmental Regulations, Inc., http://www.strongerinc.org. The STRONGER state review process involves teams representing industry, states, environmental and public interest groups reviewing state oil and gas waste management programs.

[116] Ground Water Protection Council, *State Oil and Natural Gas Regulations Designed to Protect Water Resources*, pp. 7, 39-40. STRONGER has been reviewing state programs to assess whether any revisions may be needed to address issues surrounding the use of hydraulic fracturing in oil and gas development. Hydraulic fracturing reviews have been conducted for Arkansas, Pennsylvania, Ohio, Louisiana, and North Carolina, Ohio, and Oklahoma.

II wells nationwide.[117] In contrast, the DOE Energy Information Administration reports that the number of gas producing wells in the United States increased from 302,421 in 1999 to 487,100 wells in 2009, and most new wells are fractured.[118]

EPA's annual appropriation includes funds for state grants to support state administration of various EPA programs. Since the 1980s, annual appropriations to support state UIC programs have remained essentially flat (not accounting for inflation) at roughly $10.5 million to $11 million.[119] Ten EPA regional offices and 42 states share this amount annually to administer the full UIC program, which covers 1.7 million wells (Classes I through V) nationwide.[120] The GWPC has estimated that annual UIC program funding would need to increase to $56 million to fully meet the needs of the existing UIC program.[121] The GWPC further estimated that EPA would need to provide funding at a level of $100 million annually to meet the needs for the full UIC program, including the regulation of geologic sequestration of carbon dioxide. Given the large number of wells that are fractured, UIC program oversight and enforcement costs for state agencies could be considerably higher if this process is subjected to federal UIC regulations. EPA and states would need to develop new regulatory requirements for these wells and increase staff to review applications and make permitting decisions. States and industry representatives have expressed concern that failure to provide sufficient resources would likely create permitting backlogs. For example, under UIC regulations, EPA or the primacy state must provide for a public hearing for each permit issuance, and have inspectors on site.[122] Some states impose permit fees or use other revenue-generating mechanisms, while such approaches have not been embraced in other states.

Because of the sheer number of potentially newly regulated wells, EPA (given its current resource levels) would necessarily need to rely heavily upon the states to implement this program. In 2007, the GWPC noted that states are already struggling to fully implement their UIC programs, and new requirements for hydraulic fracturing would be problematic. The GWPC cautioned that without substantial increases in funding for the UIC program:

- More states will decide to return primacy to EPA (which also would not have additional funds to implement the program).

- The overall effectiveness of UIC programs will suffer as more wells and well types are added without a concurrent addition of resources to manage them.

---

[117] U.S. Environmental Protection Agency, *FY2008 Drinking Water Factoids*, EPA 816-K-08-004, November, 2008, http://www.epa.gov/safewater/databases/pdfs/data_factoids_2008.pdf.

[118] EIA, *Natural Gas Navigator Number of Producing Gas Wells*, August 2009, http://tonto.eia.doe.gov/dnav/ng/ ng_prod_wells_s1_a.htm.

[119] Congress provided $11.84 million for FY2011 and $10.85 million for FY2012.

[120] Mike Nickolaus, Ground Water Protection Council, UIC Funding Presentation, Ground Water Protection Council 2007 Meeting January 23, 2007, http://www.gwpc.org/meetings/uic/2007/proceedings/Nickolaus_UIC07.pdf.

[121] Ground Water Protection Council, *Ground Water Report to the Nation A Call to Action*, Underground Injection Control, Ch. 9, Oklahoma City, OK, 2007, http://www.gwpc.org. This estimate preceded EPA's promulgation of new UIC regulations establishing Class VI wells for geologic sequestration of carbon dioxide and EPA's determination that production wells that use diesel must receive a Class II permit.

[122] See requirements at, for example, 40 C.F.R. 144.51(m), *Requirements prior to commencing injection*. Also, 40 C.F.R. Section 124 11 provides for public comments and requests for public hearings for UIC permits. The UIC program director is required to hold a public hearing whenever he or she finds a significant degree of public interest in a draft permit (40 C.F.R. §124.12(a)). Section 124.13 states that a comment period may need to be longer than 30 days to allow commenters time to prepare and submit comments.

---

- Decisions regarding which parts of the program to fund with limited dollars could result in actual damage to USDWs if higher risk/higher cost portions of the program are put "on the back burner."

- Negative impacts on the economy could occur as permitting times lengthen due to increased program workloads.[123]

EPA resources are also at issue. The agency would require additional technically trained staff to oversee and enforce state programs and implement the program in non-primacy states (such as New York and Pennsylvania). Should some states decide not to assume primacy for the new program, EPA's resource challenges would increase. As with states, EPA resources are stretched. For example, the agency is continuing its review and approval of various state Class V UIC programs that are being revised to implement a 1999 rulemaking. Additionally, EPA published a rule in 2010 establishing UIC requirements for the geologic sequestration of carbon dioxide.

# Studies and Research

Technical and practical questions regarding the development of the unconventional oil and gas resources remain unanswered. In 2009, U.S. Geological Survey (USGS) researchers noted that while drilling and hydraulic fracturing technologies have improved over the past several decades, "the knowledge of how this extraction might affect water resources has not kept pace."[124] Consequently, environmental regulators, oil and gas developers, and communities have faced new challenges and some uncertainties as these resources are developed. State regulations, industry practices, and technologies are evolving.

Several studies and research projects are under way related to hydraulic fracturing for the purpose of oil and gas development. In August 2009, DOE announced that it was funding nine new research projects intended to improve methods for treating, reusing, and managing water associated with natural gas development—including gas from coal beds and shale. Several of these projects address hydraulic fracturing, including projects to develop processes and technologies for pretreatment of produced brine and hydraulic fracturing flowback waters. Another project is intended to develop a new hydraulic fracturing module to assist regulators and operators in enhancing protective measures for source water and streamlining the well-permitting process.[125] Such research studies could help reduce water contamination risks associated with fracturing and reduce regulatory impacts.

In EPA's FY2010 appropriations act (P.L. 111-88), Congress directed EPA to carry out a study on the relationship between hydraulic fracturing and drinking water, using a credible approach that relies on the best available science, as well as independent sources of information.[126] EPA expects

---

[123] Mike Nickolaus, Ground Water Protection Council, UIC Funding Presentation, January 23, 2007.

[124] U.S. Geological Survey, Water Resources and Natural Gas Production from the Marcellus Shale, U.S. Department of the Interior, Fact Sheet 2009-3032, May 2009, http://pubs.usgs.gov/fs/2009/3032/pdf/FS2009-3032.pdf.

[125] U.S. Department of Energy, *DOE Projects to Advance Environmental Science and Technology  Nine Unconventional Natural Gas Projects Address Water Resource and Management Issues*, August 19, 2009. List of projects is available at

http://www.fossil.energy.gov/news/techlines/2009/09058-DOE_Selects_Natural_Gas_Projects.html.

[126] P.L. 111-88, H.Rept. 111-316:

Hydraulic Fracturing Study.—The conferees urge the Agency to carry out a study on the (continued...)

to report on the interim research results in 2012, and issue a follow-up report in 2014. EPA's Draft Hydraulic Fracturing Study Plan states that the overall purpose of the study is to understand the relationship between hydraulic fracturing and drinking water resources. Specifically, the study is designed to examine conditions that may be associated with potential contamination of drinking water sources, and to identify factors that may lead to human exposure and risks. EPA has proposed research studies that address the full lifecycle of water in hydraulic fracturing, from water acquisition and chemical mixing, through actual fracturing and to post-fracturing stages, including the management of flowback and produced water and its treatment and/or disposal.[127]

As part of the study, EPA is investigating existing reported incidents of drinking water resource contamination where hydraulic fracturing has occurred. These retrospective case studies will be used to determine the potential relationship between reported impacts and hydraulic fracturing activities. Also, prospective case studies include sampling and water resource characterization before fracturing occurs, and then evaluating any water quality or chemistry changes afterward. The EPA studies may add insight regarding the risks of hydraulic fracturing as well as specific regulatory gaps and needs.

In March 2011, President Obama announced a broad "Blueprint for a Secure Energy Future." In it, the President asked the DOE Secretary to identify steps that can be taken to improve the safety and environmental performance of shale gas production, and to develop consensus recommendations on practices to ensure the protection of public health and the environment, including water quality.[128] In November, the Secretary of Energy Advisory Board (SEAB) Shale Gas Subcommittee issued a final report, with recommendations for state and federal governments and industry. Water quality recommendations, aimed mainly at the states, include (1) adopting best practices for well construction (casing, cementing, and pressure management), (2) adopting requirements for background water quality measurements, (3) manifesting all water transfers across various locations, and (4) measuring and publicly reporting the composition of water stocks and flow throughout the fracturing and cleanup process. The SEAB also suggested that states review and modernize rules and enforcement practices.[129]

# Concluding Observations

Hydraulic fracturing bills introduced in the 112th Congress and previously have generated considerable debate. Industry and many state agencies have argued against regulation of hydraulic fracturing under the SDWA, and note a long history of the successful use of this practice in developing oil and gas resources. Industry representatives argue that additional federal regulation

---

(...continued)

relationship between hydraulic fracturing and drinking water, using a credible approach that relies on the best available science, as well as independent sources of information. The conferees expect the study to be conducted through a transparent, peer-reviewed process that will ensure the validity and accuracy of the data. The Agency shall consult with other Federal agencies as well as appropriate State and interstate regulatory agencies in carrying out the study, which should be prepared in accordance with the Agency's quality assurance principles.

[127] Information on EPA's hydraulic fracturing study is available at http://water.epa.gov/type/groundwater/uic/class2/hydraulicfracturing/index.cfm.

[128] See http://www.whitehouse.gov/sites/default/files/blueprint_secure_energy_future.pdf.

[129] U.S. Department of Energy, The Secretary of Energy Advisory Board, Shale Gas Production Subcommittee, Second Ninety Day Report—November 18, 2011, http://www.shalegas.energy.gov/.

is unnecessary and would likely slow domestic gas development and increase energy prices. At the same time, the amount of natural gas produced from unconventional and conventional formations that relies on hydraulic fracturing continues to grow. Drilling and fracturing methods and technologies have changed significantly over time as they are applied to more challenging formations, increasing markedly the amount of water and fracturing fluids involved in production operations. It is the rapidly increasing and geographically expanding use of hydraulic fracturing, along with a growing number of citizen complaints of groundwater contamination and other environmental problems attributed to this practice, that has led to calls for greater state and/or federal environmental oversight of this activity.

The central issue in the debate has concerned the need for, and potential benefits of, federal regulation of hydraulic fracturing. Pollution prevention generally, and groundwater protection in particular, is much less costly than cleanup, and where groundwater supplies are not readily replaceable, protection becomes a high priority. Environmental regulations generally involve internalizing costs associated with processes. And federal regulations generally are used to address activities found to have widespread public health and environmental risks, particularly where significant regulatory gaps and unevenness exists among the states. To the extent that a regulation is needed and is well designed and implemented, public benefits (i.e., protecting water resources) would be expected to accrue. If EPA were to regulate fracturing, the environmental benefits could be significant if the risks of contamination were significant and states were not effectively addressing those risks. Alternatively, the benefits may be small if most pollution incidents are found to be related to other oil and gas production activities, such as improper disposal of produced water or mishandling of materials on the surface. Some of these issues are not subject to SDWA authority and would not be addressed through regulation under this act.

State oil and gas and groundwater protection agencies widely support keeping responsibility for regulating hydraulic fracturing with the states. In September 2009, the GWPC approved a resolution supporting continued state regulation of hydraulic fracturing and encouraging Congress, EPA, DOE, and others to work with the states and the GWPC to evaluate the risks posed by hydraulic fracturing. The GWPC and others have expressed concern that regulation of hydraulic fracturing under the SDWA would divert compliance and enforcement resources from higher priority issues. Additionally, the IOGCC has adopted a resolution urging Congress not to remove the fracturing exemption from provisions of the SDWA, noting that the process is a temporary injection-and-recovery technique and does not fit the UIC program which EPA generally developed to address the permanent disposal of wastes.

Nonetheless, given the importance of good quality water supplies to homeowners, ranchers, and communities, and uneven regulation across the states, some continue to urge a federal solution. It could be expected that the potential impact of federal regulations on states and industry would be mitigated (and provide fewer added benefits) to the degree that states currently have effective requirements, or respond to increased development of unconventional gas and oil resources with their own revised requirements. In the past few years, a number of oil and gas producing states have revised their regulations to address changes in the industry.

Whether state or federal, regulations require adequate implementation resources to be administered effectively. The regulation of hydraulic fracturing under the SDWA could pose significant new staffing and other resource demands on EPA and the states; however, states that have compatible requirements in place to address hydraulic fracturing may not experience significant impacts.

Currently, there is little agreement as to the risks that hydraulic fracturing operations pose to underground sources of drinking water, and Congress has directed EPA to study this matter. The results of this and other studies should provide a better assessment of potential risks, and may help inform the need for additional regulation—whether at the state or federal level. EPA currently is developing permitting guidance, based on UIC Class II regulations, to help states regulate the use of diesel in hydraulic fracturing operations. The final diesel guidance may provide the clearest insight into how EPA might regulate this process broadly, if Congress authorized EPA to do so.

## Author Contact Information

Mary Tiemann
Specialist in Environmental Policy
mtiemann@crs.loc.gov, 7-5937

Adam Vann
Legislative Attorney
avann@crs.loc.gov, 7-6978